Forensic Facial Reconstruction

Forensic facial reconstruction is the reproduction of an individual's face from skeletal remains. Used when other forms of identification are very difficult or impossible, it can give a name to the dead in forensic cases or, in archaeological contexts, provide a tangible impression of real individuals from our past. This comprehensive work starts with a discussion of the importance of the face in society and the history of facial reconstruction, going on to evaluate the accuracy of modern reconstruction methods. The Manchester method of facial reconstruction and the relationships between the hard and soft tissues of the face are described in detail. Uniquely, the book also describes the methods and problems associated with reconstructing the faces of children. Collating all published facial tissue data and describing tissue variations with reference to age, sex, stature and ethnic origin, this book will be an important reference volume for all practitioners in the field.

CAROLINE WILKINSON is Unit Manager at the Unit of Art in Medicine at the University of Manchester, which has an international reputation for forensic and archaeological facial reconstruction. Dr Wilkinson's research has focused on facial tissue depths and juvenile facial reconstruction, and she is currently involved in the development of a computerised facial reconstruction system. In addition to her work as a forensic expert she has also worked extensively on reconstructions for archaeological investigations, including those for television programmes on the BBC, Channel Four and Discovery.

Forensic Facial Reconstruction

Caroline Wilkinson

University of Manchester

CAMBRIDGE
UNIVERSITY PRESS

PUBLISHED BY THE PRESS SYNDICATE OF THE UNIVERSITY OF CAMBRIDGE
The Pitt Building, Trumpington Street, Cambridge, United Kingdom

CAMBRIDGE UNIVERSITY PRESS
The Edinburgh Building, Cambridge CB2 2RU, UK
40 West 20th Street, New York, NY 10011–4211, USA
477 Williamstown Road, Port Melbourne, VIC 3207, Australia
Ruiz de Alarcón 13, 28014 Madrid, Spain
Dock House, The Waterfront, Cape Town 8001, South Africa

http://www.cambridge.org

First published 2004

Printed in the United Kingdom at the University Press, Cambridge

Typefaces Swift 10/15 pt. and Helvetica Neue Condense   *System* LaTeX 2_ε [T B]

A catalogue record for this book is available from the British Library

Library of Congress Cataloguing in Publication data
Wilkinson, Caroline, 1965–
Forensic facial reconstruction / Caroline Wilkinson.
    p.   cm.
Includes bibliographical references and index.
ISBN 0 521 82003 0 (hardback)
1. Facial reconstruction (Anthropology)   2. Forensic anthropology.   I. Title.
GN74.W55 2003
599.9–dc21   2003051244

ISBN 0 521 82003 0 hardback

To Frances.
You are my sunshine.

Contents

Foreword

The human face has fascinated artists, physicians, acquaintances, friends, lovers, and man and womankind since the fall of Adam. Some of the greatest endeavours of our species have sought to explain, describe, categorise, reproduce or represent the almost inexplicable nuances of detail that render us visually individual. Two of the artificial divisions of human activity – art and science – have invested huge effort in attempting to understand why we look as we do. We are taken through a history of the face from Aristotle to Bertillon, and we are left with a reminder of the pervasive nature of facial appearance in history, art and science. It is relatively recently that disciplines as apparently diverse as sculpture, psychology, anatomy, dentistry, criminology and forensic science have collaborated in an attempt to 'identify the unidentifiable' – to pull the rabbit out of the hat and rebuild the ravaged and damaged face to make it once more recognisable.

A moment's thought will convince the reader that to become an 'expert' in these matters, let alone to bring together in one volume the current state of the art, requires substantially more than a broad artistic or scientific training. The author has excellent credentials, since whilst trained in art, sculpture and facial reconstruction, she has applied rigorous scientific assessment to her own work and the work of others, and has produced a comprehensive work of scholarship that will be required reading for those with the most eclectic connection with the mysteries of their own faces and the faces of others.

The problems of facial reconstruction are formidable and the author recognises that to be successful a 'dedicated period of study' is essential. This field is not for the dabbling amateur. She acclaims the 'Manchester Method' of reconstruction as the one of choice and pays homage to her teacher and mentor, Richard Neave. The accuracy of the reconstructed face is dependent upon multiple factors, but not least the scaffolding on which it is built – the skull. the author's skill in her work and her ability to condense it into this wide-ranging text is no less dependent on the rigid support and expertise of those who developed this fascinating discipline.

No textbook can be fully comprehensive but in this volume are covered the importance of the face, the history of its study and particularly that of its reconstruction, the methods available, their accuracy and examples from the criminal world of how recognition leading to identification may be achieved. The book will be of great interest to all who are curious about 'the face' but especially to anatomists, psychologists, dentists, criminologists, photographers and artists. Although the author notes that 'unlikeable faces are more easily recognised [and presumably remembered] than likeable ones', this will not be true of the review of reconstructed faces covered so well in this 'likeable' book.

D. K. Whittaker
Professor of Forensic Dentistry
University of Wales College of Medicine

Acknowledgements

This book would not have been possible without the inspiration and teaching of Richard Neave, for whom I have a great deal of respect and affection. Special thanks are given to Denise Smith, Ray Evans and June Wilson at the Unit of Art in Medicine, who have encouraged, aided and ignored me at all the right moments.

I am indebted to Professor David Whittaker, Dr Robert Stoddart, Dr Valerie Hillier, Sophie Mautner, Elaine Chiang, Dr Manish Motwani, Denise Smith and Richard Neave for collaboration in the research included in this book. I would like to thank Professor Richard Helmer for the equipment and instruction, Tony Bentley and Nick Ogden from the University of Manchester for photographic services and Caroline Needham for illustration work.

A significant contribution was made by all those who gave their time to act as volunteers in my research. Special thanks to Penny Dale, Helen Corr and Susannah Reeves, whom I pestered beyond the limits of friendship.

I am very grateful to Phil Eva and Chris Rynn for the proofreading provided, and for giving up their time at such short notice.

I must also thank Mr Paul van den Hoven from the Netherlands Forensic Institute, Mr John Davies, Fern Corns, Nile Evans, Mandy Miller, Ray Evans, Stuart Bowman, Lee Cousens, Jun Midorikawa, Professor Peter Egyedi and Ted Reed for help with some of the images.

Unlimited thanks must be given to my parents, Eric and Clarice, who have never suggested that I was not capable of anything.

Finally, I must recognise the enormous amount of support, both practical and emotional, offered by my partner Phil and daughter Frances, without whom I would never have finished this book.

Introduction

Facial reconstruction is a process whereby the face of an individual is built onto the skull for the purpose of identification. The theory behind facial reconstruction is that in the same way that we all have unique faces, we all have unique skulls, and it is the small variations in the shape, form and proportions of the skull that lead to significant variations in our faces. When I first became involved in facial anthropology, I too had a great deal of difficulty believing that the amount of variation seen in the world's population of faces could also be exhibited in skulls. Even though we are all experts at facial recognition and identification, due to our innate ability to distinguish one face from another, we find it difficult to believe that the skull can provide a detailed map for the face. This must be, in part, due to our inability to distinguish one skull from another in the same way that we can distinguish one face from another. Uninitiated observers will not be able to demonstrate proportional and feature variation between skulls with ease. Since all skulls appear similar in shape and proportions to the inexperienced eye, it is assumed that the information provided by one skull must be virtually the same as that provided by another skull. However, the practised and experienced observer can demonstrate unlimited variation in shape, size, proportion and detail between skulls. I am now convinced: each skull is as individual as each face.

Facial anthropology is an unusual field made up of professionals from a wide variety of backgrounds including medical art, forensic art, dentistry, computer science, anthropology, archaeology, forensic science and anatomy. Since the field is so varied there has, in the past, been a paucity of reference books, and students have had to carry out research in all these areas in order to collate all the required information. Over the last thirty years a number of books have been published that have become giants in this field.

Such works as *The Human Skeleton in Forensic Medicine* (Krogman & Iscan, 1986), *The Face Finder* (Gerasimov, 1971), *Forensic Analysis of the Skull* (Iscan & Helmer, 1993), *Making Faces* (Prag & Neave, 1997), *In the Eye of the Beholder* (Bruce & Young, 1998), *Craniofacial Identification in Forensic Medicine* (Clement & Ranson, 1998) and *Forensic Art and Illustration* (Taylor, 2001) can be considered touchstone publications. However, there are still a number of missing links, and this book attempts to fill those gaps. This book will, for the first time, collate all the published facial tissue depth data, so that it can be easily referenced and applied to the particular ethnic group of the individual being reconstructed. In a world growing rapidly more cosmopolitan, this should enable reconstruction practitioners to apply the most appropriate set of data to the forensic investigation, and may also succeed in encouraging more researchers to widen the database for facial tissue depth studies. In addition, I have tried to build on the work produced by Prag & Neave (1997), by describing the Manchester method of facial reconstruction in great detail, to allow future practitioners to follow this technique precisely. Finally, the field of juvenile facial reconstruction will be discussed for the first time. This discipline has never been studied separately from adult facial reconstruction, and warrants a chapter to itself as it throws up challenging and interesting problems all of its own.

As humans we have an enormous fascination for faces. The first chapter of this book attempts to discuss the importance of the face as a social and biological tool, the attempts of scientists, artists and philosophers to understand the significance of the face, and how the process of facial recognition affects the field of facial reconstruction.

Chapter 2 is a review of the history of facial reconstruction, from the early attempts at skull over-modelling to the computer-generated systems developed in recent research. The pioneering work of Mikhail Gerasimov, Wilton Krogman and Richard Neave in the development of reproducible and effective facial reconstruction techniques is described.

The importance of the skull in individual identity is discussed in Chapter 3, where sex, age and racial origin determination from the skull is detailed. Additional information regarding the pathological assessment of the face is outlined for the depiction of facial trauma and disease.

Currently there is no single source for the majority of research into the relationship between the soft and hard tissues of the face. Chapter 4 attempts to cover all the research results, references, discussions and disagreements regarding the determination of feature detail from the skeletal structure. I have tried to establish rules and standards for facial reconstruction by illustrating the research under facial feature headings.

Facial tissue depth information is very important in the field of facial anthropology and although there are many papers containing research data, they have never been collated into one source for easy access for practitioners. Chapter 5 reviews the historical techniques of facial tissue depth study and describes the development of the use of medical imaging. The differences between data from different ethnic groups, sexes, ages and methods of measurement are discussed, and the chapter is illustrated with tabulated results from a wide variety of ethnic, nationality, sex, age and cultural groups.

Chapter 6 is a thorough and detailed description of the Manchester method of facial reconstruction. This chapter describes skull re-assembly, skull casting, skull preparation, facial muscle sculpture, facial modelling and case study recording. This is the most detailed, published description of this method since *Making Faces* by John Prag and Richard Neave (1997).

Since questions regarding the accuracy of facial reconstruction are constantly raised, Chapter 7 attempts to discuss the positive and negative research results relating to reproducibility and accuracy of forensic facial reconstruction. Previously published studies are examined and described in an attempt to devise the most appropriate method of assessment and to establish the level of resemblance that can be expected from the various techniques.

Finally, Chapter 8 introduces the field of juvenile facial reconstruction and discusses the problems associated with reconstructing the faces of children. The differences between sex, age and racial origin determination between juvenile and adult skulls are illustrated and the facial growth pattern from infant to adult is described. Finally the reconstruction method for children is described, and the accuracy of juvenile work is discussed, illustrated by forensic case studies.

1   The human face

'Out of the great number of faces that have been formed since the creation of the world, no two have been so exactly alike, but that the usual and common eye would discover a difference between them.'

William Hogarth (1753)

No two faces are alike, not even those of identical twins. Each face is unique. The human face is one of the most important social tools. An enormous variety of communication signals are produced using the face and it governs the expression of emotion, interest, desire and attention. We use our faces to attract, repel, scare, soothe and entertain. The face suggests details such as age, gender, culture, health and ethnic group. It is usually the first part of the body that we notice and the only part that we address. We are each capable of perceiving the smallest variation between faces and it is this ability that allows us to carry out personal recognition and identification. As adults we recognise and differentiate hundreds of faces of our family, friends, colleagues, famous people etc. This is illustrated by our ability to distinguish eventually the faces of identical twins (see Fig. 1.1). Even though the faces of identical twins are remarkably similar and they share the same genetic profile, their faces are in fact slightly different and people who know identical twins very well can distinguish them from one another. The smallest alteration in the configuration of a facial feature can substantially alter the appearance and character of the face as a whole. Landau (1989) felt that facial appearance suggests our identity, and that personality and individuality are encoded in the face. He suggested that the majority of the facial appearance is determined at

Fig. 1.1 Identical twins. Identical twins have the same genetic profile and remarkably similar faces. However, their faces will be slightly different from one another. Henry and George Cooper, boxers. With permission from Rex Features.

birth by genetic factors and although environmental factors will affect and alter aspects of facial appearance, the face develops and ages according to a predetermined schedule. From birth to death our face undergoes enormous changes but the individual identity remains throughout. At birth the face is a quarter of its adult size, and during infancy the development of the brain is precocious to the development of the face. During the first year the face will more than double in size, and throughout childhood the bones and cartilages develop and alter the proportions and shape of the face. During adulthood the face continues to change in more subtle ways, such as the appearance of wrinkles and loss of elasticity of the skin, loss of tension in the lips, loss of lustre of the eyes, rounding of the jawline, teeth loss and hair variation. Despite all these changes the individual's identity remains apparent. Our ability to recognise familiar faces is little altered by time, as illustrated by Bahrick *et al.* (1975) who studied how accurately we can recall the faces of old schoolmates. Even after 35 years the results showed that the participants recognised former classmates with 90 per cent accuracy.

The importance of the face

The significance of the face has long been a topic for speculation by philosophers, scientists and artists. Clearly the face is very important for communication and social contact, and child development research shows that the child's view of a person begins with the face. A child will first draw a person as merely a face and later will add the arms and legs radiating from the head. Finally the torso develops and the drawings may begin to appear more realistic (see Fig. 1.2). But many philosophers and scientists believed that the face could also tell us something about the personality and character of an individual. Hogarth (1753) wrote that '*the face is the index of the mind*'. The deduction of character from facial

Fig. 1.2 Drawings of a person by a child aged two (A), three (B) and four (C) years.

morphology is known as physiognomy and it has a long historical tradition. Aristotle (384–322 BC) was thought to be the first to
apply physiognomy in his book *Historia animalium*. He wrote that a
large forehead suggested slowness of movement, a small forehead
suggested fickleness, a broad forehead suggested a tendency to be
distraught and a rounded forehead suggested a quick temper. He
also described the eyebrows as indicative of softness of disposition
(straight), harshness (curving down to the nose), humour (curving
out to the temples) and jealousy (drawn in towards one another).
The ears suggested a tendency to irrelevant talk (large and protruding) and the eyes were indicative of impudence (open and star-
ing) and indecision (winking). Giambattista della Porta (1535–1615),
a renaissance advocate of physiognomy, wrote *De humana physiognomonia* (1586). He claimed that both character and facial features
were a result of man's temperament, and he divided faces into four
basic types: sanguine, phlegmatic, choleric and melancholic. Della
Porta invented and described optical instruments and, although a
very scholarly scientist, he also tended to interpret the world mag-
ically and spiritually, and referred constantly to analogies between
plants, animals and men. He suggested that men who resembled
a donkey were like that animal in temperament: timid, stupid and
nervous. Similarly, men who looked like pigs were greedy, rude
and irascible; men who looked like ravens were impudent; those
like oxen were stubborn and lazy; those like lions were coura-
geous and magnanimous; and those like an ostrich were timid,
elegant and stolid (see Fig. 1.3). Della Porta's early work in physiognomy influenced the eighteenth-century philsopher, Johann
Caspar Lavater (1741–1801), who wrote numerous books establishing
rules for the determination of personality from the morphological
assessment of the face. Lavater wrote, '*There is significance in every
part of the body; in the combined whole, therefore, is that astonishing expression which enables us to form a prompt and unerring judgment of
every object.*'

Many caricaturists have also exposed the similarities between
human and animal faces to infer personality to famous people

Fig. 1.3 Della Porta's illustration of physical appearance as indicative of personality. From *De humana physiognomia* (1586), V&A Library.

(see Fig. 1.4). Justified or not, certain facial characteristics seem to suggest certain personality traits in others and there are many recorded accounts where people are grossly misjudged from their facial features. Probably one of the most ironic stories is that of Charles Darwin (1872), who was almost rejected from the HMS Beagle by the captain because, as Darwin recounted, '...[he] doubted *whether anyone with my nose could possess sufficient energy and determination for the voyage*'. The work of Lavater and later nineteenth-century scientists such as Gall (1757–1828), and Combe (1788–1858) leads to the development of the popular pseudo-science known as phrenology or craniology, where the physiology of the brain was thought to be directly linked to the personality and character of the individual. Facial types have also been linked to personality through the influence of the endocrine system, with an adrenal influence associated with an angular face and a thyroid influence associated with a rounded face (Penry, 1939). As with all these facial-type theories, when a face does not fit into either category it can be conveniently classed as mixed.

Victorian anthropologists sought to measure intelligence from craniometrics. Broca (1824–80), a professor of clinical surgery in

Fig. 1.4 Capital. Viktor Deni (1919). Political poster depicting a capitalist with porcine characteristics.

Paris, and his followers believed that the size of the brain and facial form was related to intelligence. The academic racism of this period was routinely stated as scientific theory and Broca revealed his bigotry in an 1866 encyclopaedia article on anthropology when he stated, '*A prognathous (forward jutting) face, more or less black colour of the skin, woolly hair and intellectual and social inferiority are often associated, while more or less white skin, straight hair and an orthognathous (straight) face are the ordinary equipment of the highest groups in the human series*' (Gould, 1981). In the late nineteenth and early twentieth centuries, anthropologists sought to determine criminal characteristics from facial features. Lombroso, the famous Italian psychiatrist, studied criminals in an attempt to further his theories on the anomaly of the 'criminal type'. He measured, weighed and recorded the facial details of 383 criminals and published a book, *L'uomo delinquente* (*Criminal Man*), where he concluded that criminal types exhibited smaller cranial capacity, greater skull thickness, simplicity of cranial sutures, large jaws, facial asymmetry, a low and narrow sloping forehead, prominent brows, absence of baldness and anomalous teeth. However, when Lombroso's cranial measurements are studied they reveal no such differences between criminals and law-abiding subjects (Gould, 1981). Lombroso came under a barrage of attack from contemporary scientists and eventually retreated from his theory of atavism. Galton (1822–1911), a Victorian mathematician and celebrated cousin of Darwin, sought to construct a 'beauty map' of Britain by classifying girls that he passed in the street or elsewhere as '*attractive, indifferent or repellent*'. Predictably, for an Eastender, he found that London ranked the highest for beauty and Aberdeen ranked the lowest. The work of these anthropologists and criminologists has not stood the test of time, but although physiognomy is now considered a pseudo-science at best, it has undoubtedly influenced our visual and written depiction of facial features in relation to our interpretation of character (Jenkinson, 1997). However, not all physiognomy studies from this period are considered to be pseudo-science. The work of Bertillon (1896), the French anthropologist, remains useful. Alphonse Bertillon worked as a records

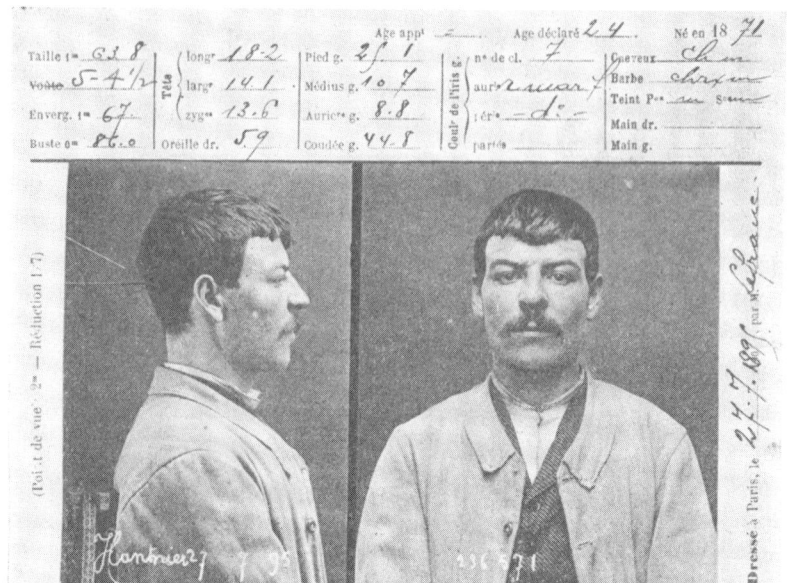

Fig. 1.5 A typical Bertillon record card. After: Personal identification (Wilder & Wentworth, 1918).

clerk at the Paris Police Force in 1877, and he was appalled to find that the records of criminals were poorly maintained. He devised a new format for identifying criminals based on face and body measurements and a morphological assessment of the person's facial and physical characteristics. The subject was photographed in frontal and profile views and eleven measurements were taken of the head, foot, arm, index finger etc. A record card was produced bearing the photographs, the measurements and a description of morphological appearance and distinguishing marks (see Fig. 1.5). Even Galton, who was one of Bertillon's greatest critics, said the system was ingenious. The Bertillon system was incorporated into the British and American police forces and although it was super-seded by the fingerprint system in the early twentieth century, the visual assessment part of the Bertillon system is still used today. The greatest flaw of the Bertillon system was the inability to en-sure uniformity in measurement, with officers producing differ-ent readings from the same subject. An additional problem was created by the effects of aging, which may alter facial and body characteristics.

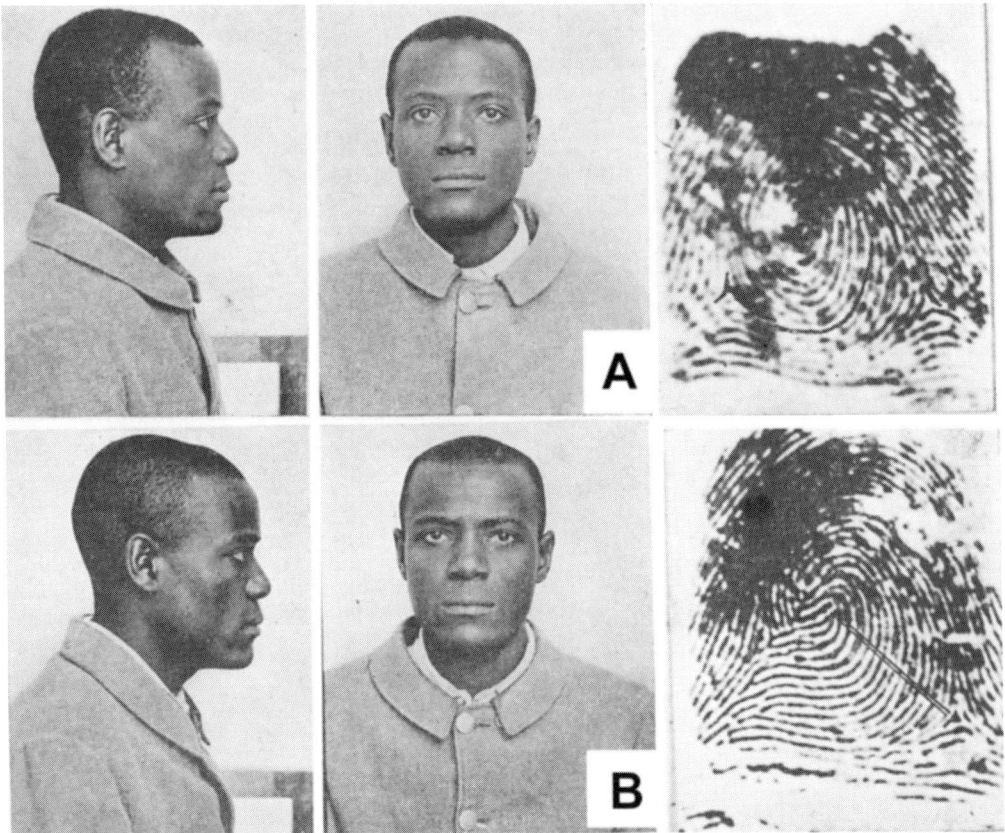

Fig. 1.6 Mug shots and fingerprints of Will West (A) and William West (B). After: Personal identification (Wilder & Wentworth, 1918).

Law enforcement myth states that 'Bertillonage' was ultimately discredited in 1903 with the infamous case of Will West (see Fig. 1.6). The story goes that Will West was placed in the United States Penitentiary at Leavenworth and when the records clerk took his measurements they matched exactly those taken from a previous offender, William West. The police suggested that Will West had lied at his trial, when he had claimed not to be a repeat offender, in order to save himself from the harsher punishments meted out to recidivists, but the convict stuck to his original story. The disagreement was cleared up when the records clerk turned the records card over and discovered that William West was still a prisoner in the penitentiary. When the two were placed side-by-side, everyone was astonished by the likeness between them. This case was truly perfect for illustrating the fallibility of the three systems of personal identification: names, photographs and Bertillon measurements. Not only did the two men have the same name, same appearance and same Bertillon measurements, but they also claimed to be totally unrelated and had different fingerprints. This single case was used to illustrate the failings of the Bertillon system and the superiority of the fingerprint system for identification. However, some authors (Cole, 2001) claim that the Will West case did not actually happen as legend suggested. Cole refuted the evidence that these events could have happened at Leavenworth Penitentiary as fingerprinting was not introduced there until 1904, and he stated that the first mention of the West case was in 1918 in a book published by Wilder and Wentworth, who used it to demonstrate that there were pairs of individuals that would be distinguished through fingerprinting, but not through anthropometry. Wilder and Wentworth did not, at this stage, claim that fingerprints of the two men were taken in 1903, or that the incident was responsible for the downfall of 'Bertillonage'. Cole suggested that the West story was concocted in order to provide a tidy story to illustrate the superiority of fingerprinting. Most importantly, however, when the Bertillon cards of Will and William West were studied, Cole found that the left foot measurements of the two men were different (28.2 cm and 27.5 cm), and outside the maximum tolerable deviation (3 mm). This suggested that strict adherence to the guidelines of 'Bertillonage' would have led to the clerk precluding them from being the same man. Cole stated that '*the move from anthropometry to fingerprinting was really a gradual process that had more to do with competing models of science and the social construction of race in America than with a single decisive case*'. The fact that the two Wests were African Americans may also be significant, and Wilder and Wentworth had previously suggested that '*all peoples of a race unlike that of the examiner are proverbially difficult to identify by the ordinary facial recognition, where the racial characteristics stand out so sharply as to obscure the finer details upon which recognition largely depends*'. Cole suggested that Wilder and Wentworth exploited the common prejudice of Americans regarding recognition of other races and the complexities of the new surnames of freed slaves, and noted that it is entirely possible that the two Wests were indeed related. So, determined advocates of anthropometry could convincingly argue that it had been incorrectly applied in the West case, and that this is probably why 'Bertillonage' has stood the test of time and is still considered a useful tool for personal identification.

Facial beauty

One of the most studied fields of philosophy and anthropology is facial beauty: how we recognise and describe it. The first recorded concepts of facial beauty and aesthetics were produced by the Ancient Egyptian culture over 4000 years ago. The ideal Egyptian face exhibited a round broad face with sloping forehead, weak brow, prominent eyes, rounded short nose, thick lips and rounded chin (see Fig. 1.7). Art and sculpture from this period portray the kings and queens with idealistic proportions (Smith, 1961). Centuries later, the Ancient Greeks established canons of facial proportion that were also employed in art and sculpture. Plato and other Greek mathematicians saw the structure of the human face as divided into three sections: hairline to eyes, eyes to upper lip and upper lip to chin. They suggested that, in the ideal face, these sections were of equal size and the width of the face was two-thirds its height (see Fig. 1.8). Plato also discussed the idea of the 'golden section', where the ratio of the whole to the larger part is the same as the ratio of the larger part to the smaller part (Cooper & Hutchinson, 1997). For example, the ratio of the whole face to the height from chin to eyes is the same as the ratio of the chin-to-eye section to the forehead. This golden section was used in many sculptures and paintings, along with other theories of proportion, such as that the ideal face is divisible by seven. Botticelli's Venus (see Fig. 1.8) fits such a proportional template with the hair occupying one-seventh, the forehead two-sevenths, the nose two-sevenths, the nasolabial distance one-seventh and the chin one-seventh. These

Fig. 1.7 The funerary mask of Tutankhamun. Egyptian Museum of Cairo. ©
PhotoSCALA, Florence.

classical ideals of beauty and proportion are still used in plastic
and reconstructive surgery (Tolleth, 1987). Other classical canons
of beauty suggest that the distance between the eyes should be one
eye-width, the features should be regular and symmetrical, the upper lip should have a Cupid's bow shape and the lower lip should
be fuller than the upper lip. In profile, the nose should have a

Fig. 1.8 The face of Botticelli's Venus. The Birth of Venus, Uffizi Gallery, Florence. The face of Venus exhibits the golden ideal of seven sections. Adapted from Liggett (1974). © PhotoSCALA, Florence.

straight dorsal ridge and the chin should be on the same vertical plane as the forehead. The Renaissance period of art in the fourteenth century re-established the theories of physical beauty from the aesthetic ideals of Ancient Greece and this can be seen in the work of Leonardo da Vinci and Michelangelo. Michelangelo's statue of David exhibits these ideal proportions (see Fig. 1.9).

However, modern anthropometrical studies of the face and head by Farkas *et al.* (1985a,b) revealed that these ideal facial canons are not routinely present in a healthy population and only occur as one of many variations. In the population, they studied the proportions of facial height canons (dividing the face into two, three

Fig. 1.9 David (detail) by Michelangelo (1504). Galleria dell'Academia, Florence. © Photo SCALA, Florence.

and four equal sections) and found the two-section canon in only 10%. The other canons could not be found in a single subject: 5% had the nasoaural canon (nose length equal to ear height), 9% had the nasoaural inclination canon (nasal inclination equals auricular inclination), 34% had the interocular canon (one eye's width between the eyes), 20% had the naso-oral canon (mouth width equal to one and a half nasal widths), and 41% had the orbitonasal canon (width of the nose equal to the interocular distance). Further studies (Farkas *et al.*, 2000; Porter & Olsen, 2001) confirmed that the frequency of valid facial canons in many different ethnic origin populations is greatly surpassed by their variation. However, one study (Wang *et al.*, 1997) found that the nasofacial canon (where the nasal width is one-quarter the width of the face) was present in the majority (52%) of Chinese faces, as opposed to only 33% of Caucasian faces.

We have preferences for facial beauty that appear to be consistent within and between different cultures. Iliffe (1960) arranged for a British national newspaper to publish twelve photographs of female faces aged between 20 and 25 years, which were selected to represent various facial types. 4300 readers responded to the request to rank the faces according to prettiness and he found that there was a common concept of facial beauty shared by men and women of all ages, from all areas of Britain and from a variety of professions. Udry (1965) repeated the experiment in the United States and his 100 000 respondents were in agreement with Iliffe's results, with the selection order differing only slightly between the two studies. These and other studies (Martin, 1964; Ford *et al.*, 1966) suggest that there is international and transcultural aesthetic agreement with regard to faces. In addition, there is evidence that even young infants agree upon attractiveness. Studies (Langlois *et al.*, 1991; Samuels *et al.*, 1994; Slater *et al.*, 2000) have shown that infants, even as old as a few days, spent longer looking at attractive adult faces (previously rated by adults as attractive) than unattractive faces.

Whilst there appears to be great concordance over which faces are attractive, there is still much debate over how we determine attractiveness (Langlois *et al.*, 1994). Our ideals of perfect beauty and proportion are reflected in our responses to those people whose facial appearance does not match up to those ideals. Mankind does not attempt to hide its preference for certain arrangements of facial features and we are all guilty of the unfair assessment of individuals according to their facial appearance. It is well documented that malformation and disfigurement can lead to many social difficulties, and studies suggest that most people react very badly in the initial stages of acquaintance with someone with a facial disfigurement. Ramsey *et al.* (1982) covertly observed people's reactions to an actress disguised with a disfiguring facial mark and then compared the reactions with those to the same actress with her normal appearance. When disfigured, the actress was approached less closely and people were more likely to approach the undisguised side. The general public takes abnormality of appearance to be indicative of abnormality of personality, and low physical attractiveness may contribute to a career of deviancy (Blair, 1937; Secord *et al.*, 1956). Attractive people are less likely to be convicted of a crime, less likely to be given a heavy sentence if convicted, more likely to succeed (in terms of income and job satisfaction) and more likely to be considered intelligent (Clifford & Bull, 1978). Peck and Peck (1970) claim that the public frequently assumes that the bearer of a severe malocclusion pattern is a slow, dull individual. Bruce and Young (1998) suggest that *'our individual identities are bound up with our faces in a way which makes facial injury particularly traumatic to deal with'*. In a study by Dion *et al.* (1972) volunteers were asked to assess personality from facial photographs. Attractive faces were rated as having more socially desirable personality characteristics, higher occupational status, greater marital and parental competence and greater social and professional happiness. Another study (Cook, 1939) asked volunteers to estimate intelligence of individuals from photographs. Their estimates were unrelated to the individual intelligence scores, but appeared consistent between volunteers. The factors affecting intelligence judgments seemed to be facial symmetry, seriousness of expression and tidiness of appearance.

Facial structure

The basic structure of the face, however, has evolved more out of biological necessity than as a tool for social interaction. The positions of the eyes and ears allow distance perception, the position of the nostrils in relation to the mouth eliminates choking, and the lips and jaw are perfectly designed for mastication, swallowing, respiration and verbal communication. The eyes are both placed at the front of the head where the fields of vision overlap to allow stereoscopic vision. This accounts for the frontally placed eyes on the faces of carnivores that need to hunt, aim with accuracy and catch their prey. Bruce and Young (1998) suggest that these biological stipulations lead to a basic face template that is surprisingly similar across many species of animal including humans, and that this facial blueprint includes two horizontally placed eyes above a central nose and mouth. It is the subtle differences between faces that create individuals. Farkas *et al.* (1985) thought that the variability of facial proportions ensures the individuality of the human face; whilst Penry (1971), the inventor of the Photofit identification system, felt that the human face is one basic recipe with a multitude of variations.

The most important factor affecting the face is the underlying skeleton. The skull is made up of 22 bones (excluding Wormian bones and ear ossicles) consisting of 14 facial bones and 8 cranial bones (see Fig. 1.10), and it is the complex and intricate shapes of these bones that provide the enormous variation between individual skulls. The skull provides the basic armature to which the hard and soft tissues of the face attach, and the shape of the face is dependent upon the skeletal structure, the hard tissues (such as cartilage) and the soft tissues (such as muscles, fat and skin). As basic face shape is similar for each person, tiny differences and sensitivity to these differences are used in identification. Facial differences and additional details such as eye colour, skin colour, hair colour and texture can suggest the sex, ethnic origin and age of a face. Enlow (1982) saw the facial skeleton, in physical terms, as a structure adapted to withstand the stresses of mastication and to protect the brain, and he felt that facial development is constrained by the biological interdependence between features, leading to only a small number of basic facial types.

The two basic skull types, dolichocephalic (long and narrow) and brachycephalic (short and wide) lead to basic face shapes (see Fig. 1.11) such as leptoprosopic (narrow and long and with protrusive features) and euryprosopic (broad and wide with flat

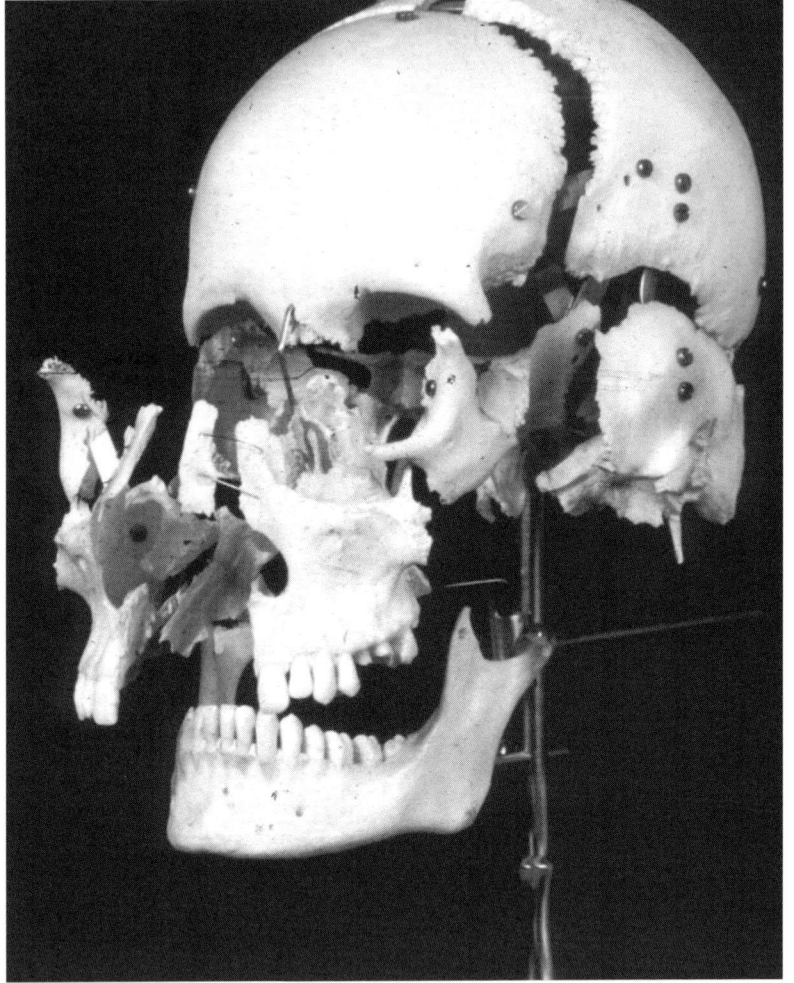

Fig. 1.10 Exploded human skull.

features). The dolichocephalic face has close and deep-set eyes, a thinner, longer and more protrusive nose that is more likely to have a high root and aquiline profile, more sloping forehead, less prominent cheekbones and retrognathic facial profile. The brachycephalic face has wide-set and exophthalmic eyes and a wider, shorter and more rounded nasal tip that is more likely to have a low root and concave profile, a more upright forehead, more prominent cheekbones and a straighter or concave profile. In fact

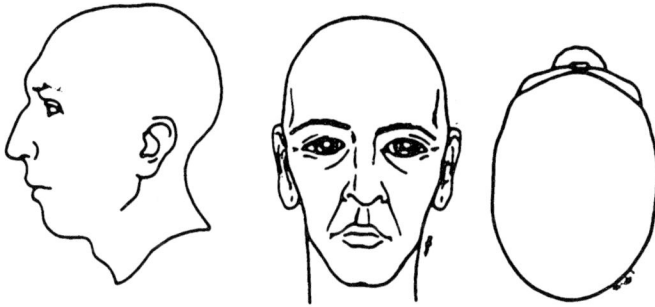

DOLICHOCEPHALIC

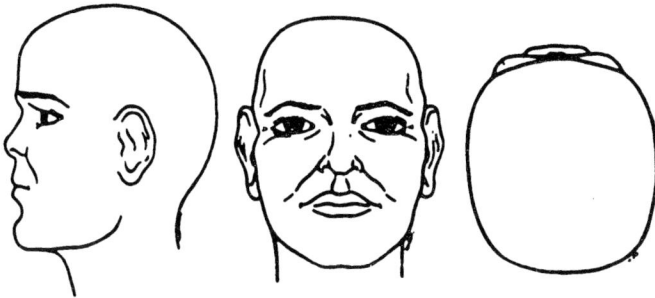

BRACHYCEPHALIC

Fig. 1.11 Facial types as described by Enlow. From *Essentials of Human Growth*, Enlow & Hans. © 1996. With permission from Elsevier Science.

it was Huxley who first introduced these classifications in 1865 (Cole, 2001). Amongst the population an intermediate facial form (mesocephalic or dinaric) may occur with intermediate facial features, but all these facial types exist in different ratios in different parts of the world. For example, Enlow (1996) suggested that the dolichocephalic type predominated in England, Scotland, Scandinavia, Northern Africa and some Near and Middle Eastern countries (e.g. Iran, India), whilst the brachycephalic type predominated in Central Europe and the Far East (Oriental).

Although it is the variation in the skeletal structure of the face that determines the characteristic feature details of individual faces, the ability to communicate is a reflection of muscular movement. Facial expression and speech are controlled by a large variety of muscles. Darwin (1872) believed that there were no specific facial muscles for expression and that all the facial muscles had primary functions in eating, looking, hearing, breathing etc. He believed that the explanation for the evolution of facial expression is as remnants of behavioural responses. The emotional responses of people from the South East Highlands of New Guinea, who were considered to be from an isolated Neolithic material culture, were studied by Ekman and Friesen (1971). Their results suggest that particular facial behaviours are universally associated with specific emotions, which may well be in agreement with Darwin's ideas.

The understanding of how the facial muscles attach and move can greatly improve the accurate depiction of faces in paintings. For centuries artists have been aware of the importance of anatomy, and many artists have used anatomical knowledge to create more realistic and lifelike expressions upon the faces of those that they paint (Hogarth, 1965). In 1844, Sir Charles Bell stated that 'anatomy is, in truth, the grammar of the language of art. The expressions, attitudes and movements of the human figure are the characters of this language.' In a remarkable painting in Basel of Christ (see Fig. 1.12) by a young Holbein, he noted that

'the painter must have been where anatomy was learned, for I am much mistaken if he has not painted from the dead body in a hospital. It is horribly true. There is here the true colour of the dead. Here is the rigid, stringy appearance of the muscles about the knee. The eyes, too, shew from whence he drew, the eyelids are open; the pupil raised and a little turned out.'

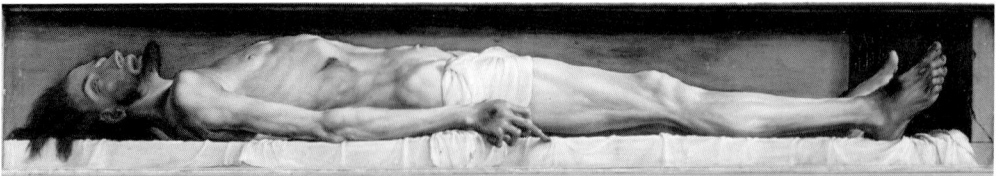

Fig. 1.12 *The Body of the Dead Christ in the Tomb* by Hans Holbein the Younger (1521). Öffentliche Kunstsammlung Basel, Kunstmuseum. Photograph: Öffentliche Kunstsammlung Basel, Martin Buhler.

Vasari (1511–74), the Italian architect and art historian, recommended for artists, in addition to the study of the antique, the frequent examination of the naked figure, of the action of the muscles and the form and play of the joints. He also advised the study of the dissected body in order to see the true position of the muscles, their classification and their insertions, and felt that from this study would come the power of realistic invention when drawing the figure. Bell (1844) went as far as to state that artists unacquainted with anatomy would have difficulty showing the form of a muscle and the complexities of design and that 'this difficulty can be traced to his ignorance of the relations and actions of the muscles'. Plastic and reconstructive surgeons have also been urged to acquire a knowledge of the skills of art in addition to anatomy, Broadbent and Mathews (1957) stating that 'we should be the greatest of all figure artists'. Many facial reconstruction artists believe, therefore, that in order to reconstruct a face upon a skull, the artist must have a deep understanding of the muscles of the face and neck, their attachments, origins and actions, and indeed Gerasimov (1971) inferred that knowledge of the topography of the facial muscles and the calculation of possible variations of them are absolutely necessary in the reconstruction of the face (see Fig. 1.13). There are many incredibly well-illustrated and well-documented anatomy books (Gray, 1973; Sobotta, 1983) and a thorough knowledge of the anatomy of the head and neck is essential in order to produce an accurate and realistic facial reconstruction. Research into facial musculature through dissection, life drawing and surgical observation will allow anatomically correct facial sculpture, and an understanding of facial nerves, vessels and glands will enable the depiction of pathological conditions, facial trauma and disease. In addition, extensive experience of the relationship between the soft and hard tissues of the face will produce well-observed and detailed facial reconstructions. The reconstructor should be able to explain why every single facial detail has been modelled and from which skeletal feature the facial information has been determined. Any facial reconstruction produced without an understanding of facial anatomy and anthropology would be at best naïve and at worst grossly inaccurate.

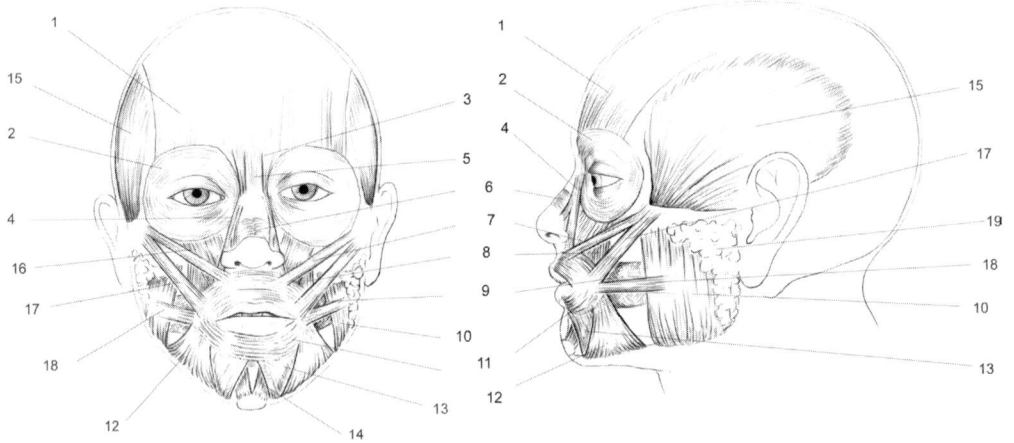

Fig. 1.13 The muscles of the face.

1 Occipitofrontalis
2 Orbicularis oculi
3 Depressor supercilii
4 Levator labii superioris alaeque nasi
5 Procerus
6 Nasalis
7 Zygomaticus minor
8 Levator anguli oris
9 Buccinator
10 Masseter
11 Orbicularis oris
12 Depressor anguli oris
13 Depressor labii inferioris
14 Mentalis
15 Temporalis
16 Levator labii superioris
17 Zygomaticus major
18 Risorius
19 Parotid gland

Facial recognition

The aim of facial reconstruction is to produce a likeness of an individual that can be recognised by someone who is familiar with the person, such as a close friend or family member. Facial recognition is one of the most widely studied areas of psychology and a great deal of important research has been produced (Harmon, 1973; Ellis *et al.*, 1979; Davies *et al.*, 1981; Brennan, 1985; Bruce & Green, 1985; Bruce & Young, 1998, 1986). Yet, despite this, it seems that it is still uncertain exactly how we recognise and process faces in our memories. We are spectacularly good at facial recognition and even babies as young as five weeks old show recognition of their mother's face. Facial recognition is a valuable talent and has been central to our social evolution. Human social interaction requires accuracy of facial recognition and Grombrich, an art historian, suggested that such recognition is automatic and based upon resemblance and biological relevance (Bruce & Young, 1998). A specific area of the brain is associated with facial recognition, illustrated graphically by the neurological condition prosopagnosia, from which a sufferer cannot recognise and identify faces (Malone *et al.*, 1982). To a prosopagnosic the face appears as a Picasso-like group of features, rather than as a whole identity. A prosopagnosic recognises familiar individuals by the voice or clothing rather than the face. Leon Harmon (1973) investigated how much information is required for facial recognition and what information is important. He designed an experiment to study the threshold of recognition by creating block portraits out of pixels based on shades of grey – a process known as spatial quantisation (see Fig. 1.14). On close inspection these images appear as a composite of squares, but viewed from a distance a face can be perceived and recognised. Pixellation can be seen when an image is displayed on a computer screen, and the more pixels that make up the image the better the reproduction of fine detail. Squinting at the image can increase recognition as this blurs the sharp lines of the pixel squares of the individual. Harmon remarked that '*it is as though the mind's eye superimposes additional detail . . . as if some kind of perceptual hysteresis prevented the image from once again dissolving into an abstract pattern of squares*'. Harmon's image of Abraham Lincoln was the most striking demonstration of how little actual facial detail is necessary for recognition of a familiar face.

Research has shown that we recognise familiar faces differently from recently viewed, but otherwise unknown, faces and it is uncertain how important each feature of the face is for recognition. Ellis *et al.* (1979) presented thirty faces of well-known celebrities, with the inner features or the outer face shape masked, to over sixty individuals. They found that famous faces could be identified using whole faces by 80% of the individuals, inner features by 50% and outer faces by 30%. When unfamiliar faces were presented there was no difference in recognition rates when the subjects were given internal or external features. They concluded that the central face is vital for recognition, not surprisingly as this

Fig. 1.14 Spatial quantisation of an image of the face of Abraham Lincoln following the technique developed by Leon Harman (1973).

region includes the eyes, nose and mouth. Many studies have been carried out on feature salience (Shepherd, 1981), and the general finding seems to be that the eyes are more important than the mouth, which is in turn more important than the nose. However, these studies were carried out using full-face images, which do not depict the shape and projection of the nose, so perhaps it is not surprising to find that the nose was less important using

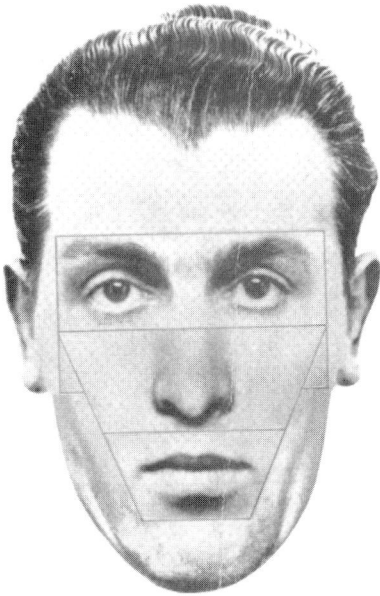

Fig. 1.15 A Photofit image using the Penry system.

this system. It has also been shown that a whole face is more than the sum of its parts (Bruce & Young, 1998) and the face is recognised more as a whole or as interrelationships between different features than as a group of individual features. Perhaps this is why it is notoriously difficult to produce good likenesses of people using the standard 'Photofit' kit made up of many different features (see Fig. 1.15). As humans we are good at recognising faces but poor at recalling them and the Photofit system allowed an eyewitness to describe the face and create a composite face from a limited variety of feature examples. The system depends on the use of over 550 photographs of actual faces that are divided into feature sections that the eyewitness can piece together like a jigsaw puzzle. However, these systems are difficult to composite into a satisfactory and recognisable likeness (Landau, 1989), and it is believed that its failure is due to inflexibility and limitations of variety. As we recognise faces as whole pictures of interrelating features rather than individual features, this system of separating faces analytically seems flawed. This view of the face as a whole can be illustrated by the

Fig. 1.16 Individual faces are difficult to recognise when part of a composite face. The composite face has been created from the top half of Gary Lineker's face and the bottom half of Paul Gascoigne's. A plausible new face emerges. Courtesy of Vicki Bruce. Figure created by Derek Carson, University of Stirling.

experiments of Young and his colleagues (1985). Their results showed that faces divided into upper and lower halves could still be recognised readily when seen without the other half. But when an incorrect lower half was added to the existing upper half, it became very difficult to identify the face to which the upper half belonged, so that upper facial appearance seems modified by the lower facial appearance (see Fig. 1.16). It is well known that inverted faces are very difficult to recognise (Bruce & Valentine, 1986) and the ability to perceive small differences in the facial proportions depends upon the faces being seen in a normal, upright orientation (see Fig. 1.17). Bruce and Young (1998) state that inverted faces appear strange and their research shows that recognition of familiar faces

Fig. 1.17 A collection of inverted famous faces. Can you recognise these famous faces upside-down? Answers left to right when right way up: Nicolas Cage, Susan Sarandon, Mira Sorvino and Kevin Spacey. With permission from Rex Features.

falls by 40 per cent when the faces are viewed upside-down. They suggest that we are insensitive to spatial relationships between inverted facial features and that it is these relationships that carry the personal identity information in a normal, upright face.

The recognition of unfamiliar faces is by no means simple. Bruce *et al.* (1999) investigated the recognition of unfamiliar target faces from high-quality video stills against arrays of ten photographs of similar individuals. The target faces were shown either in frontal or three-quarter views, whilst the array photographs were all shown in frontal view. The overall correct hit rates were only 70% (frontal – face neutral), 64% (frontal – face smiling) and 61% (three-quarter face), despite the fact that the target photograph was taken on the same day as the array photograph of the target individual. Superficial features and colouration are also important in facial recognition. Bruce *et al.* (1991) asked a number of individuals to have their faces scanned using a surface laser scanner producing a monotone image. A face pool of four similar individuals was set up and the subject was asked to pick out the individual who had been scanned. The identification rates from a face pool of photographic images

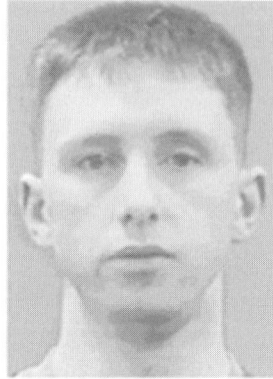

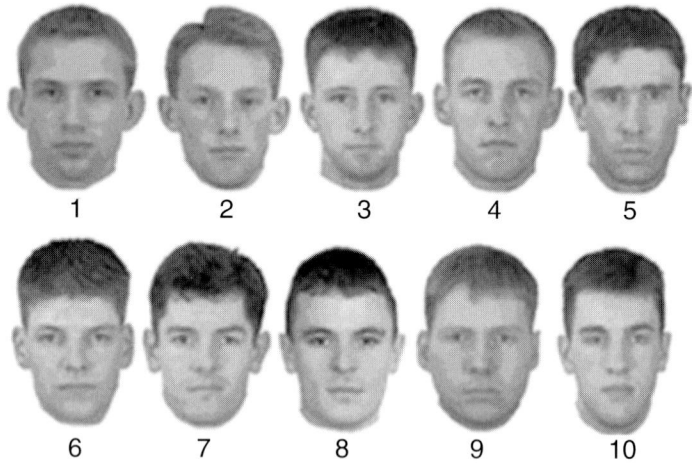

Fig. 1.18 Facial recognition study (Bruce *et al.*, 1999). A full-face photograph of a target (A) and a face pool of video images (1–10). The target individual is also image 3. Courtesy of Vicki Bruce.

were surprisingly low (26% above chance) (see Fig. 1.18). Distinctive faces may be more easily recognised than non-distinctive faces, perhaps reflected in the work of caricaturists. Caricaturists exaggerate the more unusual aspects of a face to produce a still readily recognisable drawing of an individual. Landau (1989) believes a caricature to be a visual shortcut, a potent distillation of the key ingredients that make up the distinctive features that identify a specific face. A good caricaturist recognises that the face as a whole is important in recognition but that if the distinctive features are exaggerated and the facial pattern simplified, an image that stands for something more complicated can be produced. Research by Brennan (1985) into computer-generated caricatures suggested that, when a caricature is exaggerated and recognised, it is recognised twice as rapidly as realistic line drawings of the same face, which suggests that distinctive features enhance recognition.

However, good as we are as individuals at facial recognition, we also fail to recognise individuals many times in an average day. For example, Young *et al.* (1985) asked volunteers to keep a diary of the number of times each day that difficulties or errors in person identification were made. In an eight-week period over 1000 incidents were recorded from 22 such diaries. There appear to be many factors that affect our ability to recognise faces. It is well documented (Shepherd, 1981; Kemp *et al.*, 1997) that most people find it easier to recognise, interpret and memorise faces of members of their own racial group. This is unrelated to variation within racial origin groups. A major anthropological study by Chance and Goldstein (Landau, 1989) showed that all diverse groups around the world are equally variable in facial form and features. The theory behind this 'other-race' effect is that racial groups may differ in the variability of particular features and individuals may use cues for discrimination within their own racial group, which are then not appropriate for use in other racial groups. For example, hair colour may be a distinguishing feature amongst white people (where variation is considerable), but if used for black people it may yield limited information (Shepherd, 1981). A study by Ellis *et al.* (1975) supports this theory. Twelve Black Bantu males and twelve White British males were asked to describe colour photographs of White and Black individuals and the frequency that each feature was mentioned was recorded. White subjects mentioned hair colour, hair style and eye colour more frequently and black subjects mentioned hair position, eye size, eyebrows, chin and ears more frequently. Further studies suggest that increased contact with a different racial group will improve performance of recognition in that racial group (McKelvie, 1978; Shepherd, 1981). In addition, some studies suggest that women are marginally better than men at face recognition memory (Shepherd, 1981). Sex differences have been reported as early as the first year of life (Fagan, 1972) where 5-month-old girls were shown to be superior to boys in face discrimination. Individuals differ greatly in their ability to recognise faces and many factors have been shown to affect that ability. High need-approval subjects show greater recognition of faces than low need-approval subjects, and attractive faces are correctly recognised more than faces not judged as attractive (Yarmey, 1975).

Another study by Mueller and colleagues (Landau, 1989) showed that unlikeable faces are recognised more readily than likeable faces. This research tested students using yearbook photographs. The photographs were of 80 men and 80 women, none with unusual features such as spectacles or beards. The 32 most likeable and the 32 least likeable were shown to test subjects once and then again 48 hours later. The subjects found it more difficult to remember the 'likeable' faces. Accuracy of recognition can also be impaired by the addition of spectacles, beards, hair, distinctiveness (Valentine & Bruce, 1986), angle of view (Bruce *et al.*, 1991, 1999) and the removal of surface tone and colour (Bruce *et al.*, 1991). This has been exploited over the centuries by make-up artists and individuals trying to hide their identity. It has also been suggested (Landau, 1989) that pubescent teenagers temporarily lose their keen sense of face recognition, and that this may be related to hormonal and cognitive changes associated with puberty. The combination of memory limits, ethnic group effects, sex differences, ability variation, attractiveness, and the effects of disguise and lighting mean that it is possible for the same person to look very different in different circumstances (see Fig. 1.19), and different people to appear very similar in similar circumstances (see Fig. 1.20). Although facial recognition is hard to define, one thing is certain: identity is encoded in the face and our ability to recognize and differentiate between faces is an important social skill.

Many factors must be considered when producing a forensic facial reconstruction. The forensic aim of the recreation of an individual's face is to identify the unidentifiable. Usually the

Fig. 1.19 The same man attempting to appear different. Courtesy of ABM-UK and Michael Bromby.

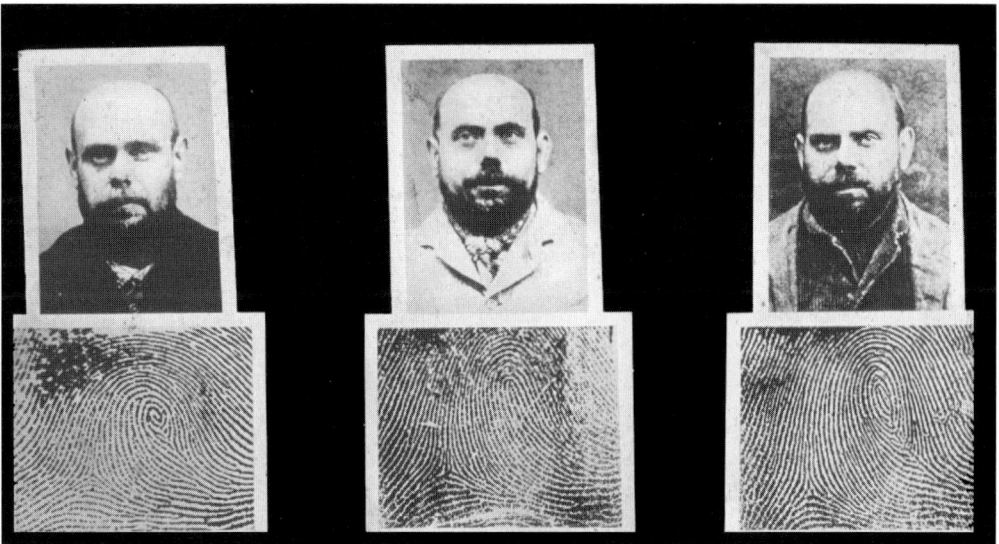

Fig. 1.20 Three men of similar facial appearance with different fingerprints. The Fox twins and father. Courtesy of Scotland Yard.

production of a facial reconstruction is a last option in a forensic investigation. The usual channels of enquiry, such as crime scene clues, missing person files and dental record assessment may have already been pursued with limited success. Facial reconstruction

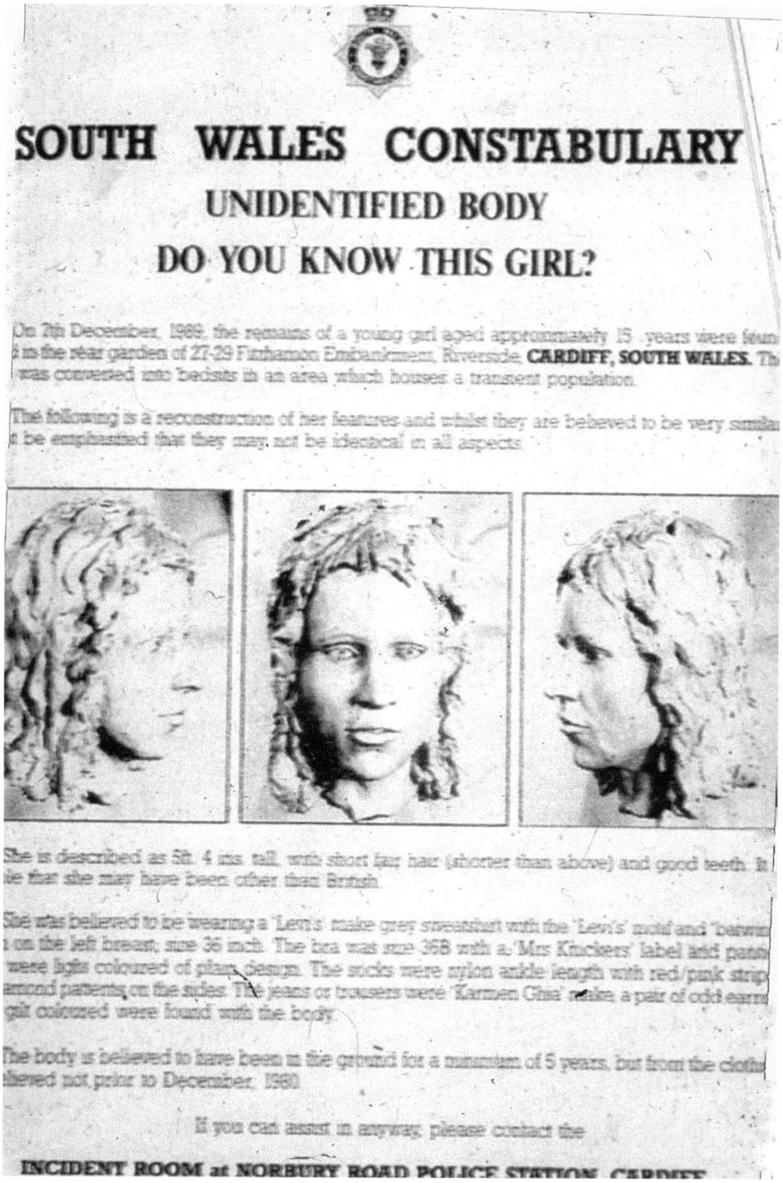

Fig. 1.21 Example of a poster used in a forensic identification investigation.

from the skeletal remains and a resulting publicity campaign may lead to recognition by a member of the public, and therefore lead to the identification of that individual. Typically such cases receive a great deal of publicity; the reconstruction may be presented on

national television and in the local and national newspapers, images of the reconstruction may be posted in the local region and,
less frequently, a publicity campaign may be mounted abroad (see
Fig. 1.21). The investigating authorities want as many people as possible to view the reconstruction, in the hope that one person who
knew the individual in life will recognise him/her. It is in their
interests to have maximum publicity for the facial reconstruction,
as the more people who view the reconstruction the more likely
it is that the individual will be identified. However, many factors
may affect whether or not an individual is recognised in such circumstances. The individual may have been missing for many years
before death or be foreign to the country of discovery. Not all those
capable of recognising the individual may watch television or read
the newspapers, and the relevant family or friends of the individual may not see the reconstruction. The age difference between
time of absence and time of death may be large and this may have
a negative effect on recognition. This effect will be more apparent
in children as a child's face may alter significantly over a short
period of time (see Fig. 1.22). With adults the facial changes related to age may not be as apparent as those seen for a child, and

Fig. 1.22 Age-related facial changes in a child. This is the same boy at five
years and 15 years of age. If this boy went missing at the age of five would
you recognise him again at the age of 15 years? Perhaps if you were his
mother, brother or sister you would recognise him.

Fig. 1.23 Age-related facial changes in adults. These are the same individuals at 20 and 60 years of age.

if you know someone well, if they are a family member or close friend, then you would expect to recognise them even after a large amount of time. However, differences in hairstyle, hair loss, subcutaneous fat distribution, skin elasticity and hair colour may have some effect on facial recognition (see Fig. 1.23).

2 The history of facial reconstruction

Facial reconstruction is the scientific art of building the face onto the skull for the purposes of individual identification. Scientific art is the application of artistic skills whilst following scientific rules. This procedure has been exercised for over a century and there exist three main techniques:

(1) The two-dimensional 2-D artistic representation of the face, usually drawn over a photograph of the skull.
(2) Three-dimensional 3-D facial reconstruction using a sculptural technique.
(3) Three-dimensional facial reconstruction using computer-generated images.

These techniques share the common principle of relating the skeletal structure to the overlying soft tissue. The technique of superimposition, which has a long history in the forensic arena, is not considered here since it requires photographic evidence of a suspect in order to connect an individual with the unidentified skull (see Iscan & Helmer, 1993; Knight & Whittaker, 1997; Whittaker et al., 1998). The artistic and sculptural reconstruction techniques have been used for recognition in forensic identification investigations worldwide, and these procedures are usually employed when the police do not have a suspect for identification.

Background

The fact that the facial reconstruction procedure exists at all is a reflection of our unlimited fascination with human faces, and this preoccupation has led to a more specific interest in the faces of

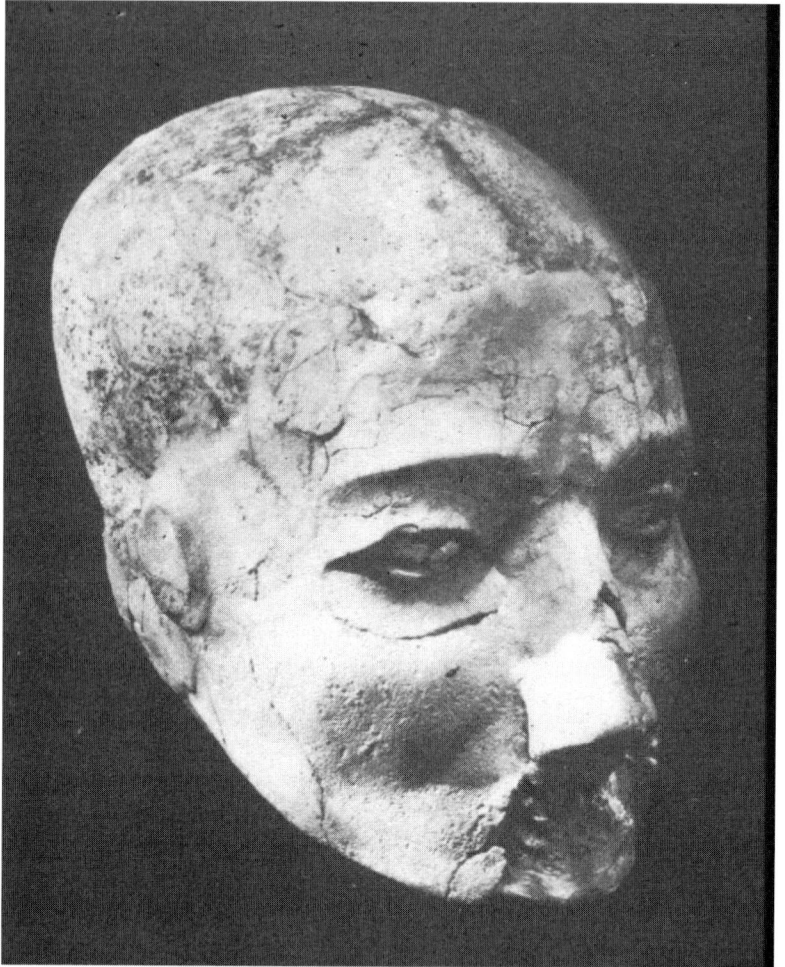

Fig. 2.1 Over-modelled skulls from Jericho (7000 BC). A number of skulls were found buried under the floor of a house in Jericho. Faces were modelled onto the calveria with local clay, and shells were set in the eye sockets to represent the eyes. The mandibles were buried separately. Courtesy of the Institute of Archaeology, University of London.

people from the past. The remains of people are constantly ex-cavated and the desire to discover how those people may have appeared has been seemingly limitless. The history of modelling a face onto a skull is extensive and there are numerous early symbolic examples (see Figs. 2.1 and 2.2). It is thought that these were not attempts to represent the faces of the individuals but

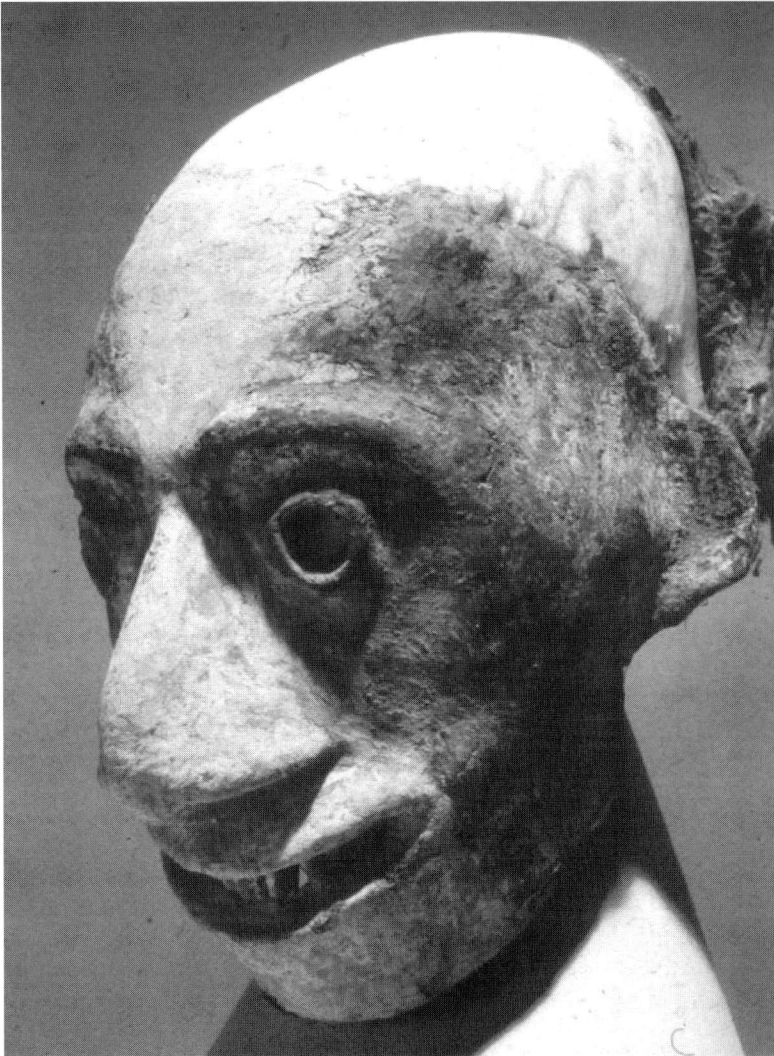

Fig. 2.2 Over-modelled skull from the New Hebridean Islands (AD 1700). A face was modelled onto the skull as a symbolic representation. Courtesy of the Manchester Museum, the University of Manchester.

merely simplistic representations, but these are certainly the first known examples of the face being sculpted directly onto the skull. Capturing the realistic features of a face has been a common theme throughout many cultures, centuries and religions. The first procedure associated with the actual appearance of the deceased

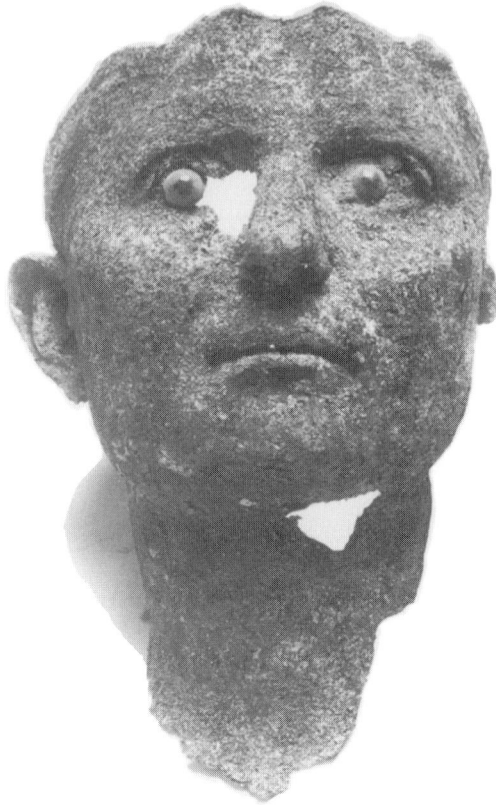

Fig. 2.3 Wax head with glass eyes from a grave near Cumae
(First century AD). With permission from the Piazza Museo, Napoli, Italy.

was the production of death masks and the first example of a death
mask was found in an Egyptian grave dating from 1370 BC. The
negative mould was usually made in plaster and the cast from the
mould was manufactured in wax (see Fig. 2.3). The second and
third centuries BC saw a major resurgence in the art of death
mask realism as part of the Roman burial ceremony, and then
a further resurgence in the medieval Christian period in order to
create bronze portraits of the deceased. In Europe, votive art (the
casting of body parts) became commonplace from the twelfth to
the eighteenth century (Schnalke, 1995). However, it was during
the Italian Renaissance that death-mask art was most appreciated,
and it became viewed with the same degree of importance as fine

art. Consequently it was the North Italian artists who introduced wax modelling to the medical world. In the fifteenth century, with the Renaissance interest in the structure and movement of the human body, these artists began to practise human dissection in order to study anatomy. Some artists, such as Andrea del Verrocchio and Michelangelo, recorded their studies in wax models, in the same way as Leonardo da Vinci recorded his studies in sketches and drawings. Following the radical transformation of anatomy teaching by the monumental work of Andreas Vesalius (1514–64) human dissections, illustrated anatomy books and teaching panels became commonplace in medical schools, and the representation of anatomy as life-size wax models offered a substitute for the cadaver. So it was, in seventeenth-century Italy, that anatomic wax modelling or 'anatomica plastica' was born, primarily from the work of Giulio Gaetano Zumbo (1656–1701), a sculptor, who created macabre scenes showing various stages of decomposition of the human body. His work exhibited both extraordinary scientific precision and an artist's sense of the inevitable decay of human beauty and power. Zumbo worked in Florence for the Grand Duke Cosimo III de' Medici between 1691 and 1694, and some of his work is housed in the Wax Anatomical Collection at La Specola. One of his most famous pieces is the head of a dead man, in wax, sculpted over a real skull (see Fig. 2.4). The facial muscle and gland structure depicted is very accurate and horrifically real, and this ranks as one of the finest examples of 3-D reconstruction.

As a result of Zumbo's work, 'anatomica plastica' spread throughout the rest of Europe, and Abraham Chovet (1704–90) was one of its greatest proponents in England. Chovet's portfolio included an écorché of a man's head depicting the muscles of the pharynx, larynx and tongue and the related nerves and vessels. Interestingly, in 1733, Chovet was accepted as a member of the 'Worshipful Company of Barber Surgeons of London', an association that still supports and collaborates with the Medical Artists' Association of Great Britain. The art of 'anatomica plastica' was further developed by the work of Ercole Lelli (1702–66) whose work can be viewed in the Anatomical Museum in Bologna. Lelli created anatomical

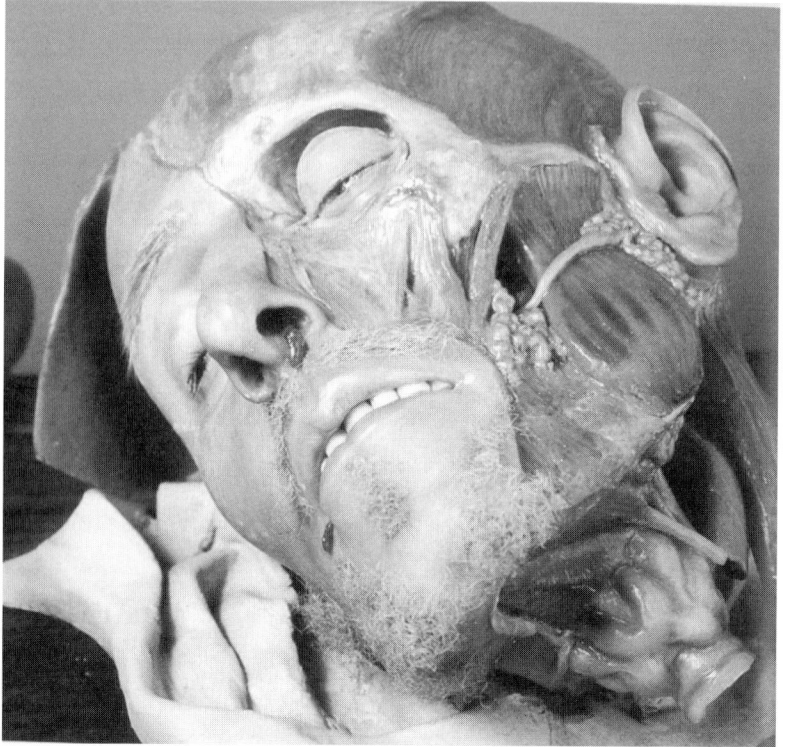

Fig. 2.4 Testa dello Zumbo. Anatomical wax model by Giulio Gaetano Zumbo, Genoa (1695–1700).

masterpieces by modelling muscles onto full skeletons, for use in medical teaching. Although we assume that these artists were less concerned with facial appearance than anatomical detail, they pioneered the development of scientific art and were the first sculptors to realise that the skeleton is the ideal armature onto which to build the musculature and the body. As many philosophers (Galen, second century AD), anatomists (Krogman & Iscan, 1986) and artists (Prag & Neave, 1997) have stated, the skull is the perfect armature for the face, and although the 'anatomica plastica' artists did not intend to create a good likeness of the individual, it is clear from viewing the models that they must resemble the individuals. These artists can be credited with pioneering the theory behind facial reconstruction: that it is the shape and proportions of the skull that provide the structure and form of the musculature, which in turn

defines the proportions and morphology of the face. The same relationship can be seen when the opposite method is performed: dissection of the face to reveal the muscles and skull. This is illustrated
in the accessible dissections produced by plastination, as developed
by Von Hagens (1979), and his exhibition of 'Bodyworlds' includes
many dissections that clearly depict the facial appearance of the
individuals.

Early attempts at facial reconstruction

The first people to become interested in facial reconstruction as
an academic exercise were anatomists. It was common practice
for skulls of famous people to be authenticated by comparison
with portraits and sculptures. The anatomist Welcker compared the
skull of Raphael with a self-portrait and the skull of Kant with his
death mask, to demonstrate that the skulls were indeed Raphael's
and Kant's. Welcker used outline drawings of the skull and the
death mask in precise orthogonal perspective and then attempted
to superimpose the outlines, whilst taking into account the soft
tissue layer. Another anatomist, Tandler (1909), followed Welcker's
method to identify the skull of the famous composer Haydn, prior
to its incarceration in the Bergkirche mausoleum erected in honour
of the composer. It was perhaps inevitable that facial reconstruction would be attempted as authentication, and the skulls of some
historic figures were reconstructed and compared with portraits,
sculptures and death masks. The famous German anatomist, His
(1895), took measurements of facial tissue from a small number of
cadavers and, using this data, modelled a bust onto a plaster cast of
the skull of the composer Johann Sebastian Bach (see Fig. 2.5). The
final reconstruction was then compared with portraits and busts
of Bach, with favourable results. Following the same method, the
skull of Dante was reconstructed by Kollman in 1898 and authenticated. These two anatomists employed sculptors to produce the
3-D facial reconstructions; Sefner worked with His, and Buchly
worked with Kollman. Sefner also worked alone to establish the

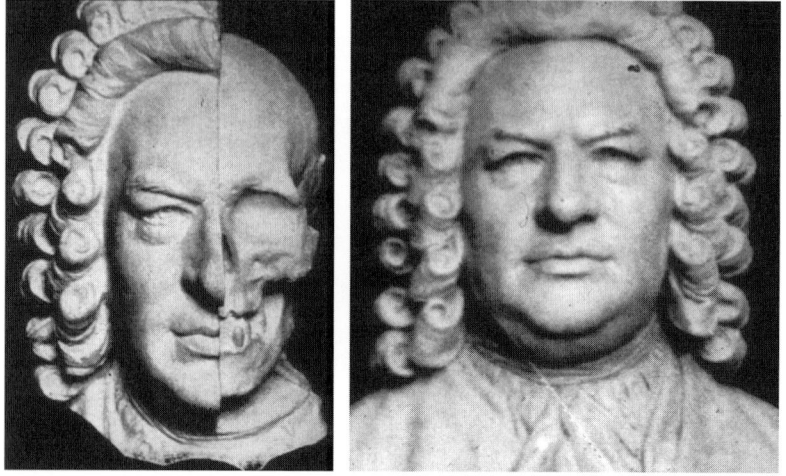

Fig. 2.5 Facial reconstruction of Johann Sebastian Bach by Wilheim His. Bericht an den Rath der Stadt Leipzig (Leipzig, 1895).

proportional connection between the skull and the face by taking the skull of Bach and attempting to sculpt onto it the face of another famous composer, Handel (Gerasimov, 1971). Although he did succeed, it was only by disregarding the proportions, size and shape of the skull; and the resulting face had areas with no tissue over the bone and other areas with grossly over-estimated tissue. Kollman and Buchly (1899) also reconstructed the face of a Stone-Age woman from Auvenier, France. This is considered to be the first real scientific reconstruction, in that Kollman used the thickness of soft tissues from hundreds of women from that area and produced a technical plan, which was then enlivened by the artistic skill of Buchly (see Fig. 2.6). The anthropologist, Gerasimov (1971) stated that *'Kollman's work can rank as one of the most remarkable achievements in the history of science-based reconstructions of heads and faces from the skulls.'*

Other early attempts at facial reconstruction were performed on prehistoric skulls and ancient archaeological specimens. In 1900, the anatomist, Merkel, was assisted by the sculptor Eichler, and together they carried out the reconstruction of an ancient Saxon. The head of a Neanderthal man from a cave at Le Moustier, France, was reconstructed by the anatomist, Solger, in 1910, and a series

Fig. 2.6 Facial reconstruction of a Stone-Age woman from Auvenier (Kollman & Buchly, 1899). *Archiv fur Anthropologie* 25 (1989) 337. By permission of the British Library (PP. 3974.d pg337).

of portraits of Early Man were reconstructed by Rutot, a Belgian anthropologist, and an artist known as Masquet. A well-preserved skull of a Neanderthal was found at La Chapelle-aux-Saints in France in 1908, and many anthropologists from countries across the world as diverse as America, Russia and Poland attempted its reconstruction. McGregor (1926) of Columbia University, a student of human palaeontology, was probably the first of a few workers to employ a half-face reconstruction technique on his Prehistoric Man skulls. Virchow (1912) recommended this technique in order to give the viewer a better idea of the relationship between the finished face and the underlying bony structure.

The obvious move from this research was to use the reconstructive technique to recreate the faces of unidentified individuals from forensic remains. In Europe at this time, many people were using a photographic superimposition technique for use in legal identification. Significant developments occurred in 1935, with a particular case involving a Lancastrian GP and the mysterious disappearance of his wife. Dr Ruxton claimed that his wife had left him for another man, but two weeks later two dismembered bodies were found in Glasgow. Police recovered two human heads and over seventy body parts wrapped in newspaper. The newspaper was from a special edition dated 15 September and distributed only in the area where the Ruxtons lived. Dr Buck Ruxton had killed his wife and her maid and removed the eyes, noses, lips, skin and teeth to avoid identification of the bodies. The police suspected that the maid and the wife were the victims and a photographic superimposition of the faces of the two women with photographs of the skulls was carried out (see Fig. 2.7) by Glaister and Brash (1937). Known objects in the photographs (e.g. a tiara and a picket fence) were used to enlarge the faces to life-size. The identification of Mary Rogers and Mrs Ruxton was considered to be successful. This case was the first documented use of this technique in a medicolegal identification case.

The development of techniques

The Russian anthropologist, Gerasimov (1971), was attempting to recreate the appearance of deceased individuals (see Fig. 2.8). In 1920, Gerasimov began studying forensic medicine under the supervision of Professor Grigoriev at Irkutsk University, and in 1935

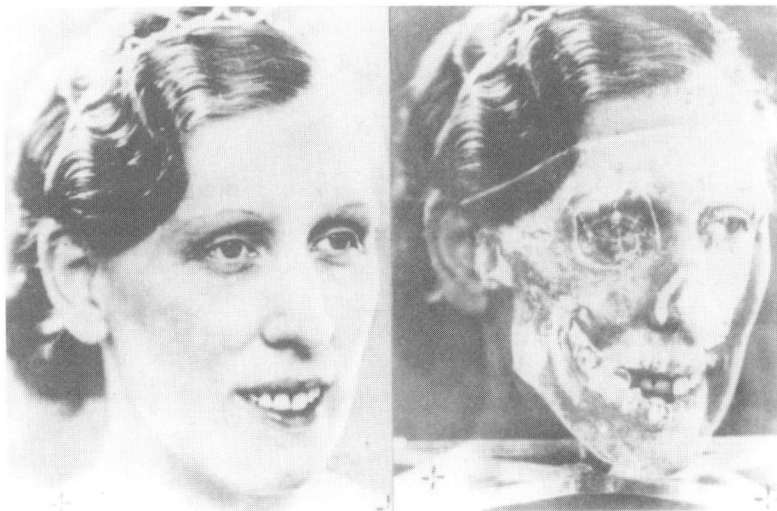

Fig. 2.7 Skull-to-face superimposition. Photograph of murder victim, Mrs Isabella Ruxton, and comparison of her face by superimposition of an image of her skull. Reprinted from *Medico-legal Aspects of the Ruxton Case*; Glaister and Brash; © 1937. With permission from Elsevier Ltd.

he modelled the head of a man and then compared the reconstruction with photographs of the man in the last few years of his life. Gerasimov was convinced that his technique could produce a good likeness of an individual, and he developed an anatomical approach that involved modelling each facial muscle onto the skull one by one. He then covered the muscle structure with a thin layer of clay to represent the skin and to create the finished face. Gerasimov was a stickler for anatomical accuracy and suggested that problems arise when investigators do not pay enough attention to muscle structure, and the interdependence of the form of the face and the peculiarities of the skull. Gerasimov split his method into two parts; reproduction of the head and then the modelling of the facial mask. The initial phase involved the reconstruction of the masticatory and neck muscles, and Gerasimov stated that '*there is no doubt that the masticatory muscles can be accurately reconstructed. They are highly individual in size, volume and shape so that their form can be in each particular case determined from the skull.*' However, Gerasimov stated that the second phase, the modelling

Fig. 2.8 Mikhail Gerasimov. From *The Face Finder* by M. Gerasimov, 1971, published by Hutchinson. Courtesy of the Laboratory of Anthropological Reconstruction, Moscow. Used by permission of The Random House Group Limited.

of the facial mask, requires special training and long experience. Against the general consensus of contemporary anatomists and anthropologists, Gerasimov suggested that details of the nose could be determined from the nasal bones, the brow and upper jaw. He also determined the mouth form from the teeth and maxillary bones; the eyes from the nasal root, orbital bones and tear ducts; and the ears from the mastoid processes, the ramus of the mandible and the auditory meatus. Gerasimov reported amazing success with his technique in the 150 forensic cases and numerous historical cases in which he was involved, but does not appear to have thoroughly documented his results. Some theorists (Tyrell et al., 1997) have cast aspersions on the honesty of Gerasimov's work, and although some of his work is poorly recorded, a number of his accuracy studies are well documented (see Chapter 7). Although he is most famous for his 3-D work, Gerasimov also produced many facial drawings from skulls, both as preparatory drawings for his sculptures and as reconstructions in their own right (Taylor, 2001). These drawings were produced without tissue depth indicators and Gerasimov shows great skill as an artist and great anatomical knowledge. Since this early work there has been a great deal of facial anthropology research in Russia led by Lebedinskaya, Balueva and Veselovskaya who carried on Gerasimov's work following his death in 1979 (Lebedinskaya et al., 1993).

McGregor, at Columbia University, was the first person to carry out facial reconstruction in the United States, and from 1915 onwards his faces of Prehistoric Man, modelled on skull casts, were a feature of the Natural History Museum in New York. However, it was probably Wilder (1912), a pioneer in forensic anthropology, who brought the attention of the North Americans to the European method of facial reconstruction, and he reconstructed the faces of many Native American skulls. Wilder (1912), with McGregor, described many valuable tips and guidelines for facial reconstruction. However, facial reconstruction was not really taken up in the United States until 1946 when the anthropologist, Wilton Krogman, took a serious look at the procedure and, with the aid of the sculptors McCue and Frost, carried out studies into the accuracy of the technique. For his first study in 1946, Krogman selected a cadaver head and photographed it before it was defleshed. The skull was then handed over to the sculptor, McCue, who produced a facial reconstruction using tissue depth data appropriate to the sex and racial origin of the individual. She was also provided with age determination. The resulting reconstruction was compared to the photograph of the cadaver and it was concluded that the reconstruction had a good resemblance to the individual, and that this

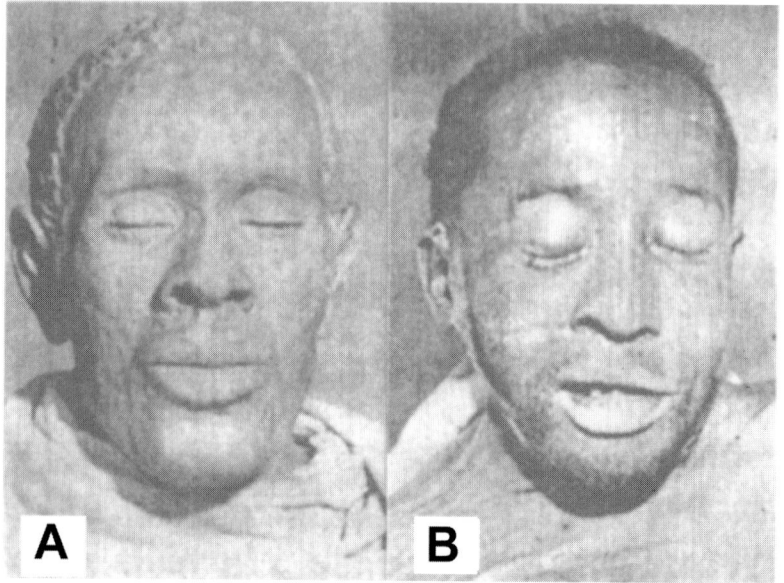

Fig. 2.9 Experiment in facial reconstruction by Krogman and McCue (1946). Facial reconstruction (A) and subject (B). Courtesy of the Federal Bureau of Investigation.

technique could be a useful tool for forensic identification (see Fig. 2.9). Krogman continued to work with many artists and often provided them with outline drawings of skulls in frontal and lateral views on which the artists would produce 2-D facial reconstructions (Taylor, 2001). In 1975, Krogman worked with a New Jersey State Police artist, Homa, using this method to reconstruct the face of a decomposed body, and the case was solved following this collaboration.

What has become known as the American 3-D method was developed from the work of Krogman by the forensic artist Betty Pat Gatliff and the physical anthropologist, Clyde Snow (Snow *et al.*, 1970), who first worked together on the investigation into the identity of a Native American man (see Fig. 2.10). Gatliff reconstructed only half of the face in this particular case and created a full face using mirror image photography. Subsequently she came to realise the importance of facial asymmetry

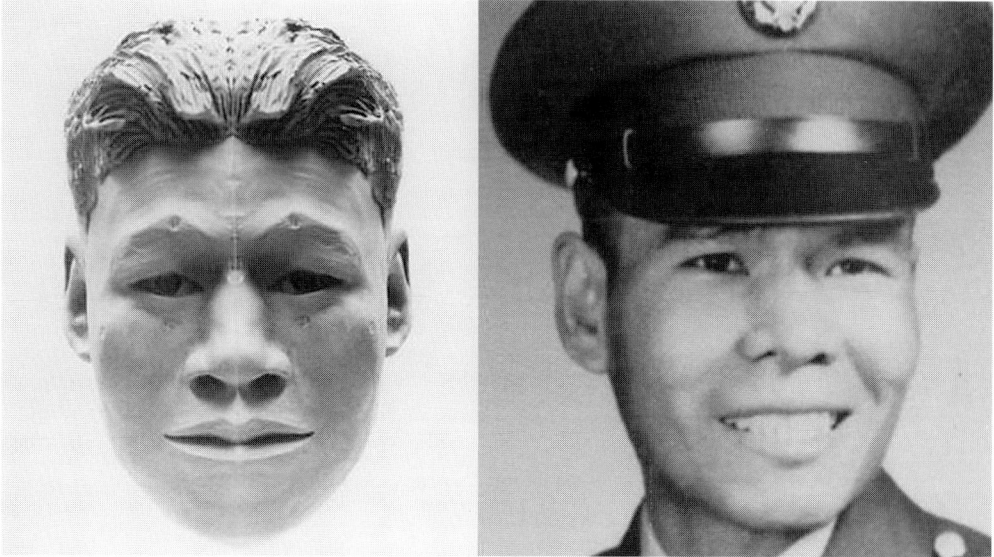

Fig. 2.10 The facial reconstruction of a Native American man by Gatliff in 1967. Facial reconstruction (left) and subject (right). Courtesy of Karen Taylor (2001).

for facial appearance, and reconstructed the whole face in future cases. Gatliff and Snow carried out a further accuracy test of the 3-D facial reconstruction technique (see Chapter 7), the results of which suggested that it could be a useful tool in forensic identification investigations. The American method employs average tissue thickness data from a variety of tables relating to different ages, ethnic groups and sexes. The skull is mounted on a stand in the Frankfurt plane. The most appropriate set of tissue thickness data is selected and cylinders of vinyl eraser are cut to the appropriate thickness and glued to the surface of the skull at the appropriate anatomical points (see Fig. 2.11A). Using modelling clay or plasticine, these markers are connected by strips (see Fig. 2.11B) to create a rough contour map of the surface of the face. The remaining open spaces are filled to form the features of the face (see Fig. 2.11C). Taylor (2001) splits the reconstruction method into two phases: the technical phase that involves information collection, skull preparation, tissue depth application and

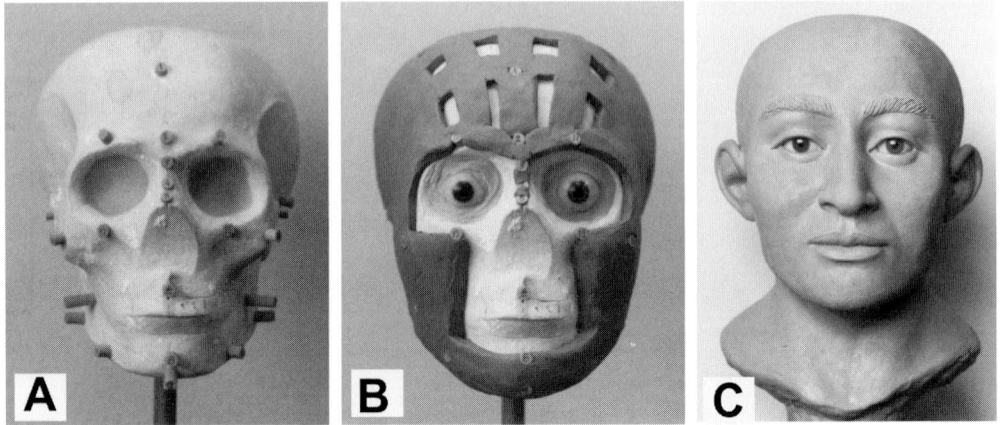

Fig. 2.11 The American method of facial reconstruction. (Snow *et al.*, 1970). A = tissue depth markers attached to the surface of the skull. B = plasticine strips join the tissue depth markers. C = facial features modelled between the plasticine strips. Courtesy of Betty Pat Gatliff.

facial contour production, and the artistic phase that involves the sculpture of the facial features and finishing of the head. All the facial features are based on the details determined from the original skull, using the 'rules of thumb' described by Krogman.

Many other anthropologists worked with artists to produce 2-D facial reconstructions. Dr. J. Lawrence Angel, of the Smithsonian Institute, worked with the police artist, Don Cherry, on a skeletal case from Columbia, and the technique was discussed in the FBI Law Enforcement Bulletin (Angel & Cherry, 1977). Angel and Cherry clearly collaborated closely during the production of this 2-D reconstruction, and the anthropologist set the parameters within which the artist worked (Taylor, 2001). The victim was positively identified. Angel continued to produce 2-D reconstructions, both with artists and on his own, following the rules set up by Krogman. Photographs of the skull were overlaid with tracing paper, onto which outlines and details of the head and face were drawn, guided by the appropriate tissue depth measurements. However, the 2-D facial reconstruction technique was really established in the United States by the anthropologist, Caldwell: a protégé of

Angel, who developed her technique by interpreting the guidelines set up by Krogman and the principles of facial feature development taught by Gatliff (Caldwell, 1981). Caldwell suggested the use of life-size frontal and lateral views of the skull onto which the tissue depth data could be added, followed by the drawing of the face. During the 1970s and 1980s many artists produced 2-D facial reconstructions following Caldwell's methodology and some artists produced muscle-by-muscle illustrations that more closely followed Gerasimov's reconstruction technique (Taylor, 2001). More recently Karen Taylor, a protégée of Gatliff, reconsidered the 2-D facial reconstruction method and suggested attaching the tissue depth markers to the skull prior to photographing, so that all the tissue depth points could be incorporated (Taylor, 2001) with the camera performing the necessary foreshortening (see Fig. 2.12). Taylor's method was further refined by the addition of guidelines regarding lateral facial features produced by George (1987). A very detailed description of the methodology for the 2-D and 3-D American methods can be found in Taylor (2001). There are many facial reconstruction artists currently working in the United States and they have a good level of success with recognition from identification. This work is an integral part of the identification procedure; and wigs, coloured eyes, skin colour, facial expression and personal belongings (watches, glasses etc.) may be employed in the reconstruction. Gatliff has also produced numerous archaeological 3-D facial reconstructions, including Leanderthal Lady (a late prehistoric female from Leander, Texas) and Tutankhamun.

In Britain, interest in facial identification came mainly from the anthropologists using superimposition techniques. Pearson (1926; 1928), for example, studied the authenticity of the skulls of Lord Darnley and George Buchanan. There have been some attempts at facial reconstruction within Europe over the last fifty years, most notably by Helmer (1984) in Germany, and Neave (Prag & Neave, 1997) in Britain. Helmer followed the American method, but used the more permanent material of modelling wax to produce his reconstructions (see Fig. 2.13). Neave, however,

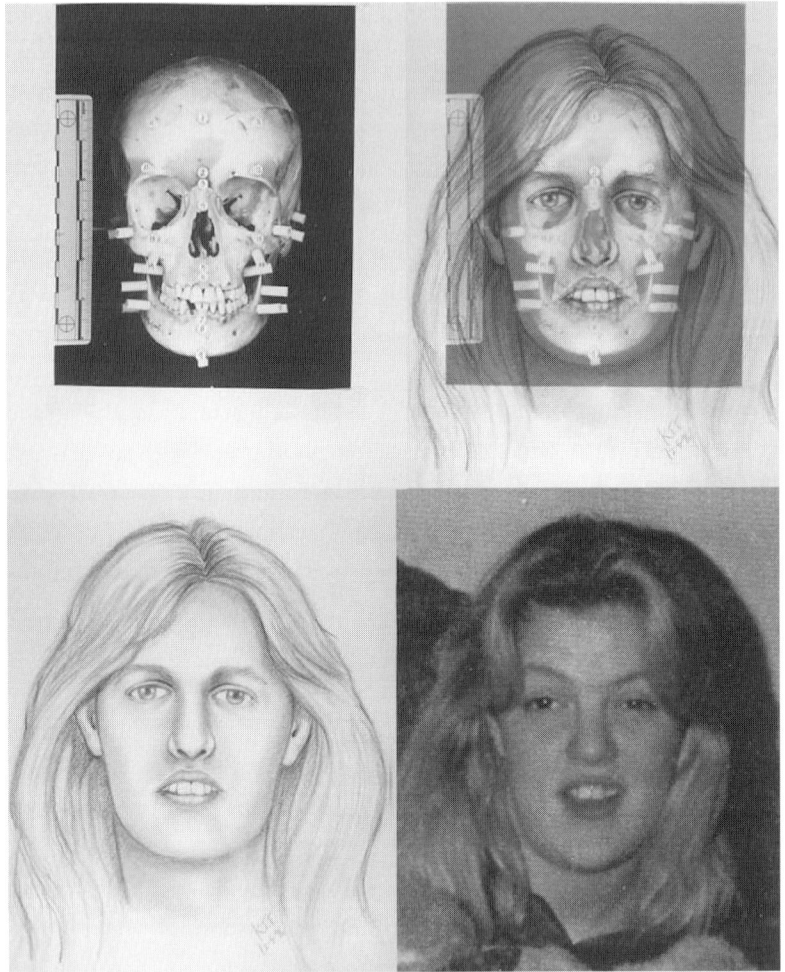

Fig. 2.12 Karen Taylor's method of 2-D facial reconstruction. Skull (top left), 2-D reconstruction overlaying image of the skull (top right), finished 2-D facial reconstruction (bottom left), and identified individual (bottom right). Courtesy of Karen Taylor.

developed a new technique, incorporating the Russian and American methods, to create a thorough and successful procedure (see Fig. 2.14), which has been taken up by many other facial reconstruction practitioners worldwide (Taylor & Angel (1998) in Australia; Hill *et al.* (1993, 1996) in Scotland). Neave became interested in facial reconstruction through the Manchester Mummy Team at the

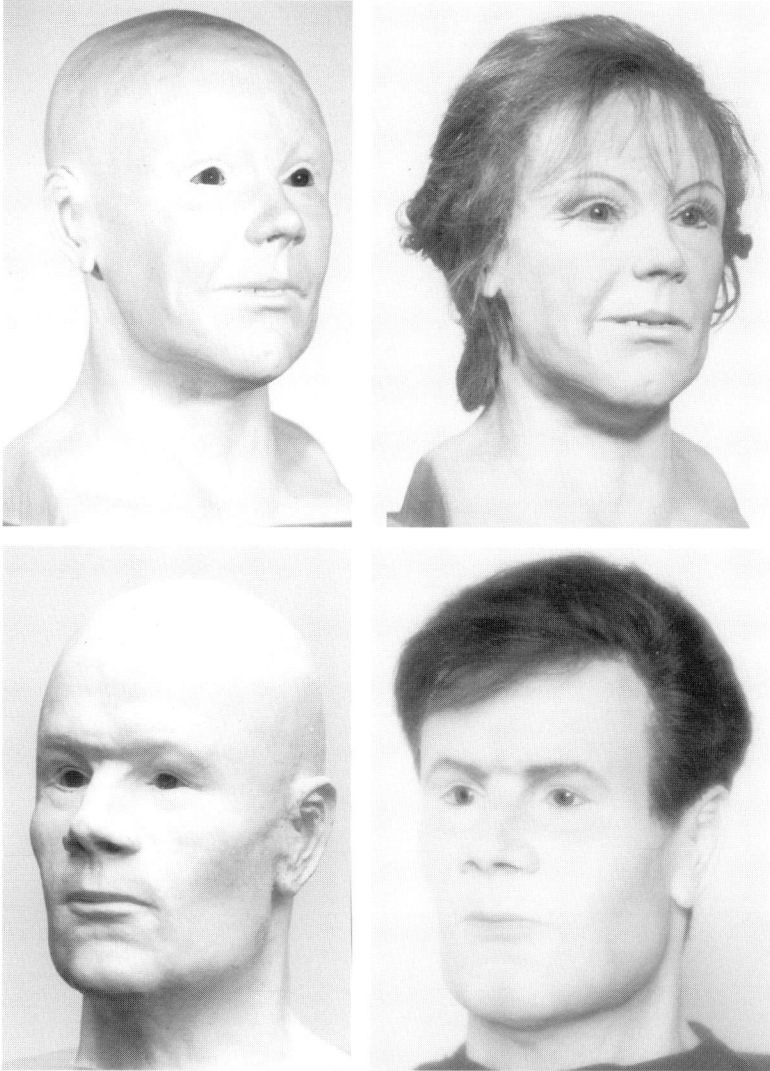

Fig. 2.13 Examples of facial reconstruction by Professor Richard Helmer. Courtesy of Richard Helmer.

University of Manchester, which was responsible for the forensic investigation of numerous Egyptian mummies housed at the Manchester Museum (Prag & Neave, 1997). The Manchester Mummy Team, an interdisciplinary group lead by Dr Rosalie David, was set up in 1970, and Neave was asked to reconstruct the faces of The Two

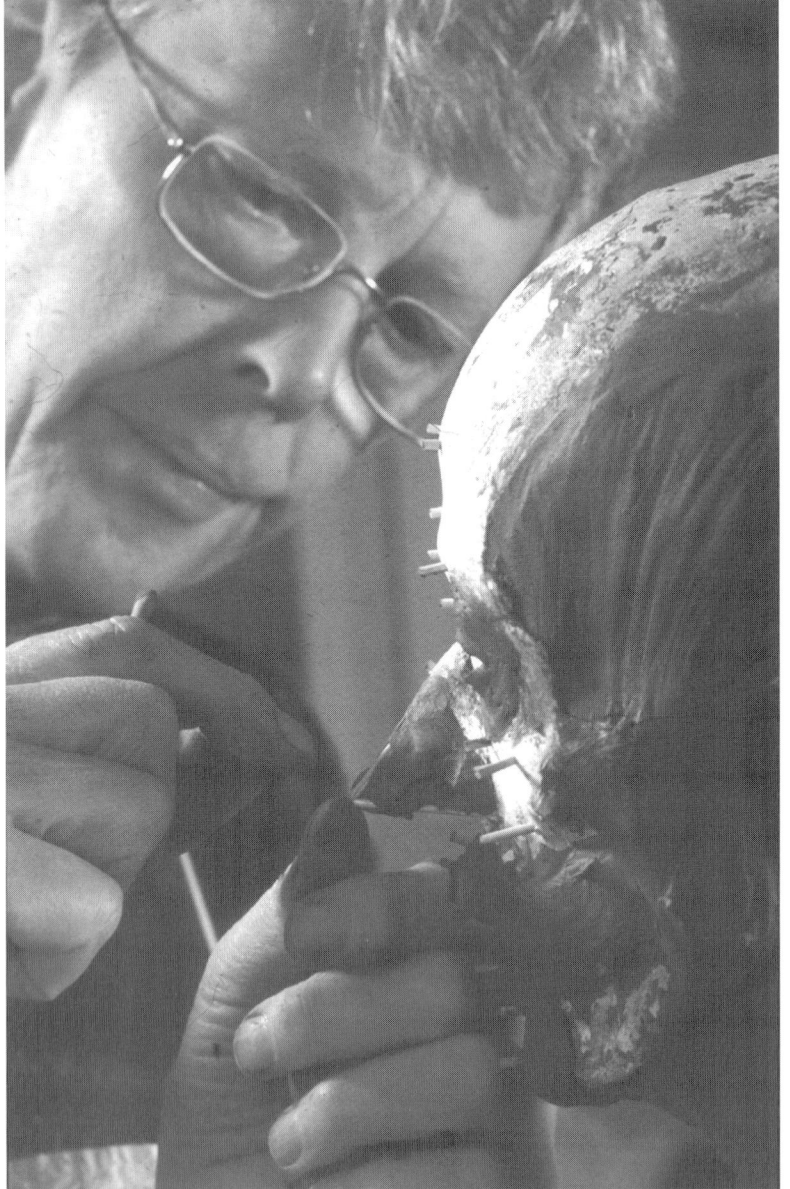

Fig. 2.14 Richard Neave.

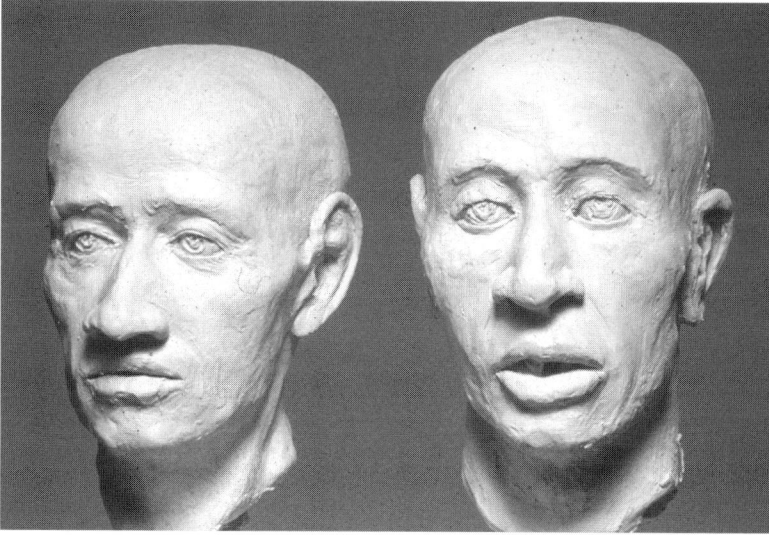

Fig. 2.15 The facial reconstructions of The Two Brothers by Neave, 1973. The Two Brothers (1900 BC) are Egyptian Mummies housed at the Manchester Museum, UK. They are thought to be half-brothers, and Nekht-Ankh (left) was 60 years old with a feminine, brachycephalic face, and Knum-Nakht (right) was 45 years old with a skull that exhibited strong Black African characteristics.

Brothers, a pair of twelfth-Dynasty Egyptians in 1973. Neave cast the skulls of The Two Brothers and facial reconstructions were produced following a rather simple and undeveloped method (see Fig. 2.15). Neave stated that *'whilst a great deal of attention was paid to the areas of muscle insertion and their probable effect upon the face, we put very little effort into developing the muscle groups themselves'*. Neave used the tissue depth data produced by Kollman and Buchly (1898) from cadavers of White Europeans for these reconstructions.

Following the success of these Ancient Egyptian investigations, Neave decided to carry out accuracy research into his facial reconstruction method. He applied his technique to four skulls taken from cadavers in the Department of Anatomy at the University of Manchester. The cadavers were photographed prior to the study and then reconstructions were produced following a slightly more developed method than that used for The Two Brothers. Neave claimed that he then correctly linked the resultant four

reconstructions to the photographs of the individuals, that the reconstructions were all different from one another, and that several
professional colleagues judged the reconstructions to be 'uncanny'
resemblances of the individuals depicted in the photographs. Over
the next 25 years, Neave developed and refined this method of
reconstruction. He employed the same sets of mean tissue depths
as the American method, but rather than basing the technique
on these limited points, he followed Gerasimov's anatomical
approach, whilst using the tissue thickness points as guides when
laying the skin layer over the muscle structure (see Fig. 2.16).
Neave worked exclusively with skull casts and sculptors' clay and
often built the neck structure onto the facial reconstructions to
give the finished head a more balanced appearance. Guidelines set
up by Gatliff (1984), Krogman and Iscan (1986) and George (1987)
were also employed to interpret the facial features. Neave and his
team were involved in approximately 20 forensic investigations
(see Fig. 2.17), with a 75 per cent success rate, and a great many
archaeological investigations. Neave's pioneering work included
Lindow Man (an Iron-Age Bog Body from Cheshire), Philip II of
Macedon (the Father of Alexander the Great) (see Fig. 2.18) and
King Midas (Prag & Neave, 1997).

The combination method of facial reconstruction, as developed
by Neave, appears to be the most accurate technique. The technique
uses all the skeletal detail of the skull to establish facial detail
and form, and relies on the tissue thickness data as a guide to
the soft tissue depth. It seems sensible to build the anatomical
detail of the muscle structure, since it is the shape and form of
the skull that inevitably influence the form and structure of the
face. Although the tissue depth data are very important, it must
be noted that these are only mean sets of tissue thickness and, as
such, cannot take into account the individuality of each skull and,
therefore, each face. The Manchester method involves the study
of facial anatomy, expression, anthropometry, anthropology and
the relationship between the soft and hard tissues of the face, and
this requires a dedicated period of training and study. The ultimate
aim of facial reconstruction is to recreate an in vivo countenance
of an individual that sufficiently resembles the deceased person. In

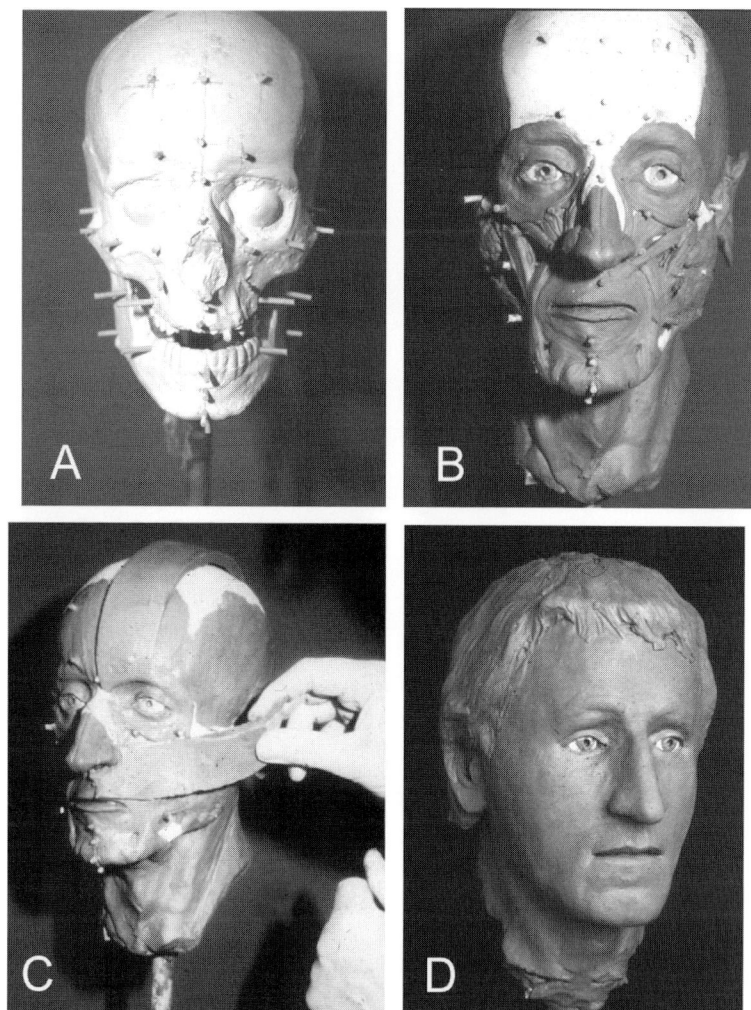

Fig. 2.16 The Manchester method of facial reconstruction. A = wooden pegs attached to the plaster skull. B = facial muscles modelled in clay onto the skull copy. C = clay skin layer placed over the facial musculature. D = finished reconstruction.

forensic situations it may contribute to their recognition and lead to positive identification via other evidence. In addition, academic study and research can attempt to increase the level of accuracy of the facial reconstruction technique, so that we can aim to reach the highest levels of resemblance when reconstructing the face of an individual.

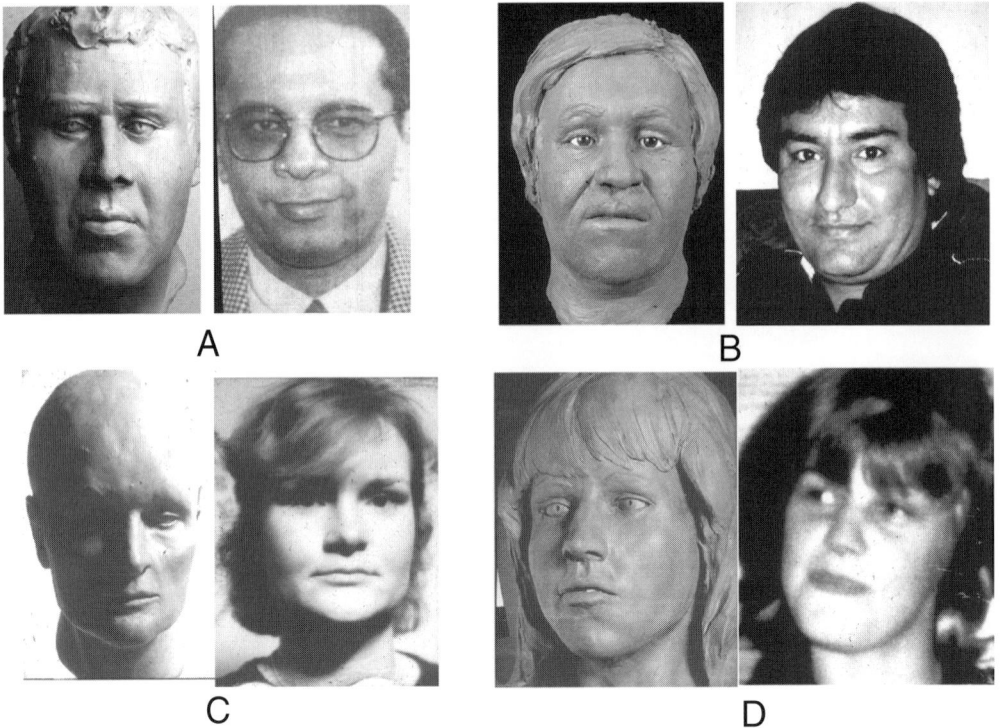

Fig. 2.17 Forensic facial reconstruction investigations from the University of Manchester. Four investigations (A–D) showing the reconstruction (left) and identified individual (right).

Computerised facial reconstruction

Over the last decade various systems have been developed to produce a facial reconstruction using computer software. The aim of these computer systems is to increase the levels of flexibility, efficiency and speed. One of the problems associated with manual reconstruction techniques is the effect of the differing knowledge and abilities of individual reconstructors. Individual variation may lead to inconsistencies in the manual technique and the resulting reconstruction may only be as accurate as the skill and experience of the reconstructor allows. The rationale behind the computer techniques is to remove the perceived subjectivity introduced by each reconstructor. The ideal computer system would involve three steps:

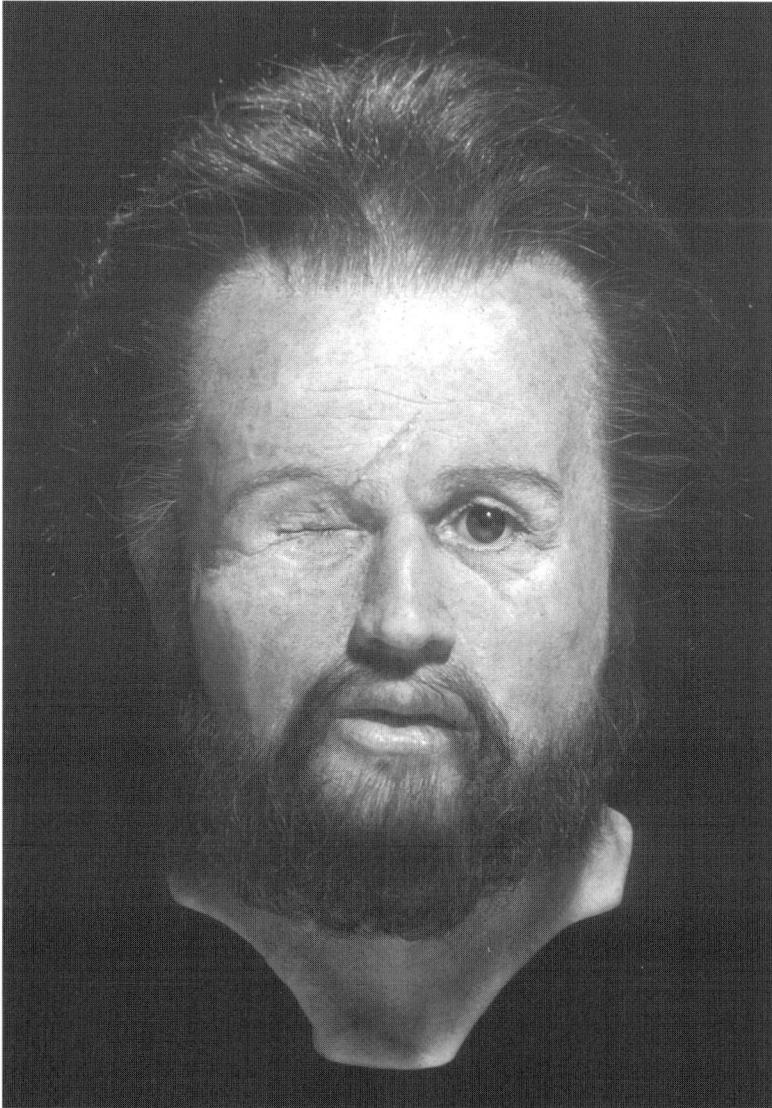

Fig. 2.18 Facial reconstruction of Philip II of Macedon (Prag & Neave, 1997).

(1) The collection of information from the unidentified skull by scanning equipment.
(2) The addition of characteristic details of the unidentified individual such as age, sex, ethnic group and stature.
(3) The production of the facial reconstruction.

The final stage of this process would ideally involve the pressing of a button and the immediate production of the face. The first development of a computer technique for forensic purposes was carried out by Moss and his colleagues (Arridge *et al.*, 1985; Moss *et al.*, 1987) at University College London (UCL) in Britain and was based upon a system used for cranial reconstructive surgery. Initially, a large number of surface co-ordinates of the skull were measured and the preferred method was an automated laser line scanner and video camera, which produced digitised surface data from the skull. This system was developed at UCL for 3-D surface data acquisition of the human face, has limited manual intervention and is subject to minimal human error. The video camera obliquely views a laser line projected onto the face/skull. The subject/skull is rotated a full 360° and the video camera is interfaced with a computer which is capable of recording 20 000 surface co-ordinates with a resolution of 0.5 mm in 30 seconds (see Fig. 2.19). Computed tomography scanning can also be used to collect skull/face data. However, this procedure delivers a high radiation dose, as the distance between each scan must be 2 mm or less, and numerous of scans may be necessary for each head. The laser-scanned data from the skull are then displayed as a fully shaded 3-D surface and the operator chooses a number of sites over the skull where the facial thickness depths appropriate to this skull are placed onto the skull surface (Vanezis *et al.*, 2000). This process relies upon the skill of the operator at placing the landmarks. An average face is then chosen dependent upon the sex, age and racial group of the skull and this face is 'morphed' to fit the new skull. Additional facial features are then added to the face, such as open eyes, hair, facial hair etc. The face is created with the eyes closed, as the laser scanning method of facial data collection cannot be carried out with open eyes (see Fig. 2.20). Vanezis *et al.* (2000) stated that *'because of the complexity and diversity inherent in the morphology and spatial relationships of facial features, it is likely that, despite our best efforts, the prediction of the true morphology of soft tissue features such as ears, nose, lips and eyes, will remain largely speculative – at least for the foreseeable future'.* A single-blind test of a known skull was carried

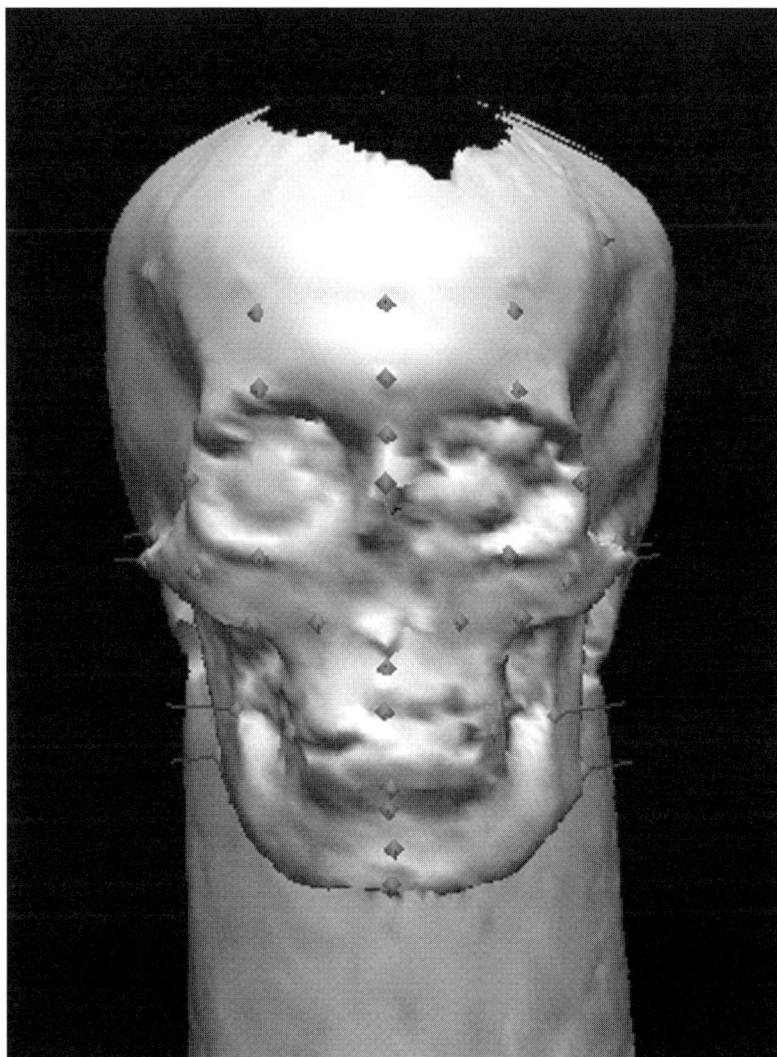

Fig. 2.19 Laser scan image of a skull. Courtesy of Maria Vanezis, University of Glasgow.

out to compare the manual technique and the computer technique (Vanezis *et al.*, 1989). The resulting reconstructions were compared with a photograph of the individual to whom the skull belonged. The results showed that both techniques can provide a useful tool for identification, although the manual technique produced a more realistic face and one which appeared more recognisable.

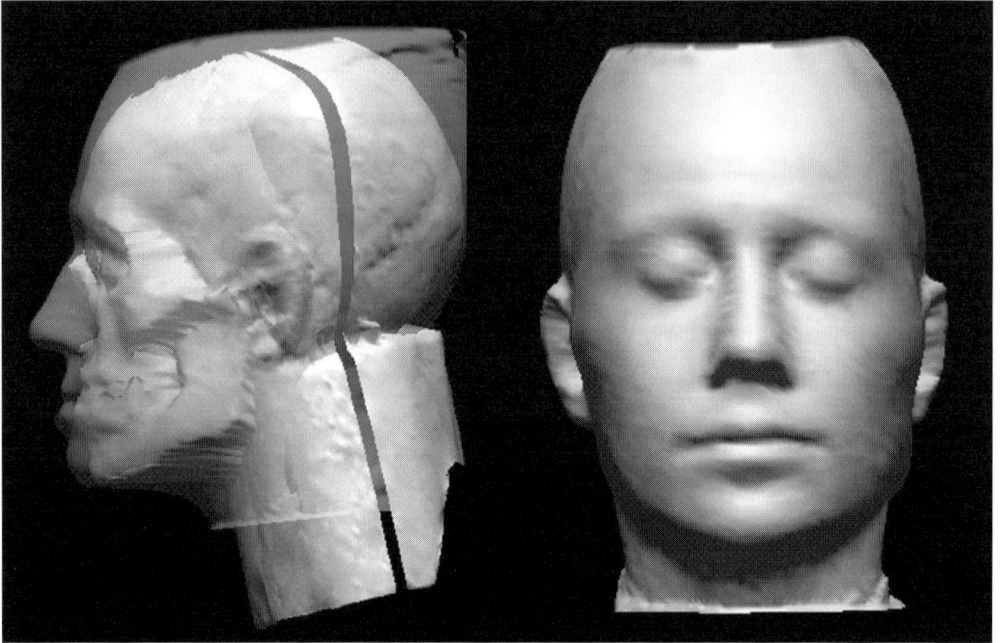

Fig. 2.20 Computer-generated facial reconstruction. Courtesy of Maria Vanezis, University of Glasgow.

The researchers stated that the manual technique suffered in that it was time-consuming, and later modifications were more difficult. They stated that the problems with the computer technique were the limited library of facial features and the over-reliance upon small numbers of facial tissue depth data. Another system by Ubelaker and O'Donnell (1992) manipulated an image of a face to fit the proportions of the skull under investigation. A video camera took an image of the unidentified skull with the tissue depth pegs already attached, and facial components were added from a database. The system used photocomposing and retouching capabilities, and is fast, easily manipulated and many variations can be stored. Ubelaker and O'Donnell stated that, '*We do not mean to imply that it is superior in all ways to clay three-dimensional reproduction. In contrast, I believe that in the hands of a skilled and experienced sculptor the three-dimensional approach offers a superior opportunity to reproduce the fine contours of the face.*'

Other similar systems use average face information from a database of photographic sections of facial components. Evenhouse *et al.* (1992) created a system that uses 22 cephalometric and craniometric landmarks, of which eight points define the perimeter shape of the soft tissue face and the remaining points determine scale, translation and rotation of the features. An average face is created on a template face, and this average face is then fitted to the unidentified skull using the landmark locations. These systems create multiple variations for each skull, but they impose a very specific set of facial characteristics onto the facial reconstruction. These faces include information, which is not determined from the skull, such as hair colour, eye colour, skin colour and texture. This end result may confuse any viewer in relation to recognition, and a more general facial reconstruction may actually trigger a memory. It is possible that the viewer may recognise a less specific reconstruction, rather than a photograph-like image of a person where estimated information is included.

Some researchers (Michael & Chen, 1996; Nelson & Michael, 1998) described the Volume Deformation System of computer facial reconstruction. The system started with a sample Magnetic Resonance Imaging (MRI) scan of a head and skull and deformed the head to fit an unidentified scanned skull. The MRI scan collected a matrix of measurements, which allowed a large number of tissue depths to be produced, although these measurements have the disadvantage of being collected from a prone face, which is affected by gravity. The chosen head was taken from a database that matched the unidentified individual with respect to age, sex and cranial form. The system was very fast and produced a face within 15 minutes from the scanning process. However, this process was largely untested. The greatest limitation of the procedure was that it used a face that the end result would ultimately resemble and therefore was only as close to the real face as the sample face. Nelson and Michael stated that the drawbacks of this system were the difficulties of reproducing the features related to the skull, but that storage was easy, the process was highly repeatable and a rerun was rapid. In addition, they stated that it would be possible to scan separate pieces of the skull and re-assemble them on screen. However, since the intact skull is a delicate balance of tension and form, this technique would require some level of testing before accuracy is accepted. Many other computerised facial reconstruction systems have been developed, with varying degrees of success (Evison *et al.*, 1999).

All the computer systems rely upon limited information for face creation and therefore their validity may be open to question. In addition, rather than removing the subjectivity of the reconstructor, these methods rely upon manual intervention for the choice of features, moulding of the features to the face, or blending of details. In addition, none of these computer methods appears to use as much skull information as the Manchester 3-D method, since the computer techniques use tissue depth data and general determinants only, rather than the more extensive information, which can be determined from the bony details. As Vanezis *et al.* (2000) state, the problems of assessing and improving the reliability of facial reconstructions need to be addressed and further research carried out.

The manual method endeavours to interpret the contours between landmarks rather than laying a face over the landmarks. Since these subtle variations in the skull are the very details that characterise our faces, the ideal technique needs to account for these idiosyncrasies. A computer system that interprets the muscle structure of the face and can be altered sculpturally would appear to be the most accurate way of developing facial reconstruction in the future. Indeed, instead of removing the subjectivity of the reconstructor, it would seem wise to harness these talents and expertise and develop a tool that can help to reduce the errors of the individual.

3 The skull

Traditionally the skull is the single most studied bone in physical anthropology, and has a complex form that develops under the influences of growth, tension and maturation. The skull is made up of 22 bones (excluding Wormian bones and ear ossicles) consisting of 14 facial bones and 8 cranial bones. It is the most complex part of the skeleton and is of major importance for physical anthropology. Knowledge of common cranial terminology is useful when assessing the skull.

The *skull* is the entire skeletal framework of the head.

The *mandible* is the lower jaw.

The *cranium* is the skull without the mandible.

The *calvaria* is the cranium without the face.

The *splanchnocranium* is the facial skeleton.

The *neurocranium* is the brain case.

The *Frankfurt Plane* is reached when a horizontal line passes through the inferior border of the orbit (*orbitale*) and the *external auditory meatus* (*porion*) on both sides of the skull (see Fig. 3.1).

Figures 3.2 and 3.3 illustrate the position and name of each of the bones of the skull and describe common cranial terms.

There are many personal identification details that can be determined from the skull. The three most important factors are the sex, age and racial origin of the person, and without these determinants it would be nearly impossible to identify the individual. A complete and undamaged skeleton can be assessed with an extremely high level of accuracy for sex (98 per cent), have the age estimated to within five years and be assigned to one of three major racial origin groups (Caucasian, Negroid and Mongoloid). However,

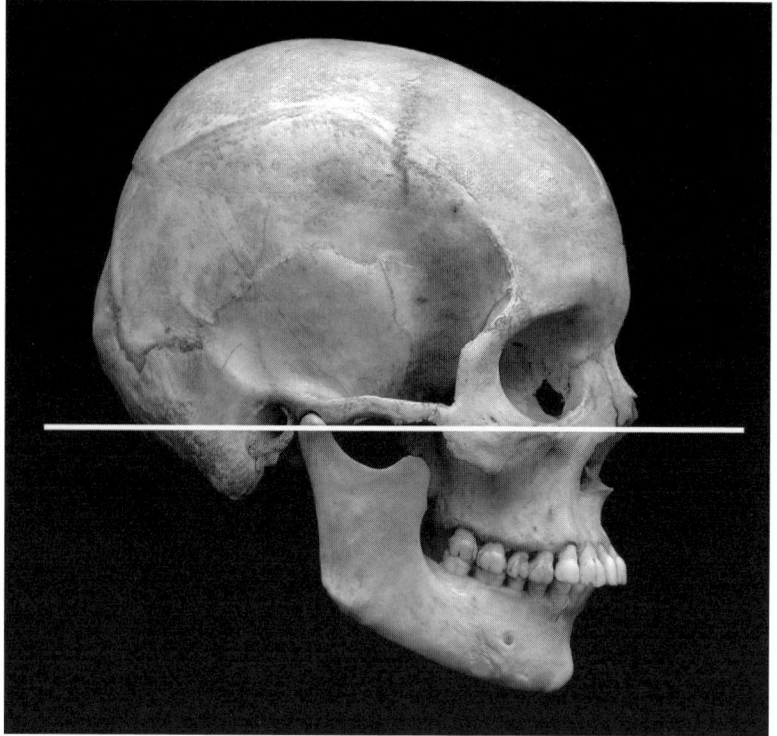

Fig. 3.1 Alignment of the Frankfurt Plane.

the accuracy of any sex, age and racial origin determination decreases appreciably if only the skull, or part of it, is available. When examining skeletal remains the forensic anthropologist tries to arrive at opinions that are really no more than classifications within the population at large (Stewart, 1979). These determinants do not identify the individual, but narrow the population for investigation. In addition to these general traits, there are individual traits that may lead to identification, such as healed fractures, dental anomalies etc., which are unlikely to be matched with any other individual.

Figures 3.4 and 3.5 describe common cranial landmarks.

Age determination

The assessment of age from the skull is primarily concerned with the teeth. There are many detailed and excellent papers discussing

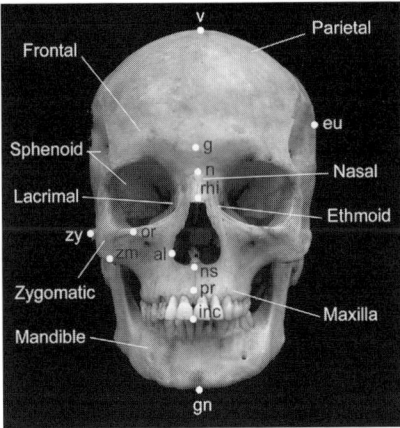

Fig. 3.2 The bones of the skull – frontal view.

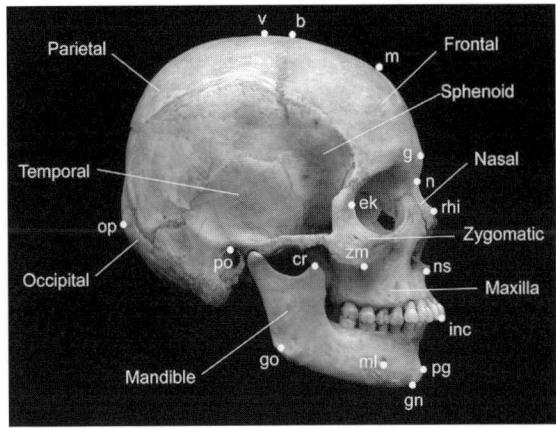

Fig. 3.3 The bones of the skull – lateral view.

al *alare* is the most lateral point on the margin of the nasal aperture.

b *bregma* is the cranial point where the coronal and sagittal sutures intersect.

cr *coronion* is the point at the tip of the coronoid process of the mandible.

ek *ectoconchion* is the most lateral point on the orbital margin.

eu *euryon* is the cranial point where the cranial breadth is greatest.

g *glabella* is the most anterior midline point on the frontal bone above the frontonasal suture.

gn *gnathion* is the most inferior midline point on the mandible.

go *gonion* is the point at the centre of the mandibular angle.

inc *incision* is the point where the central incisors meet on the occlusal line.

l *lambda* is the cranial midline point where the sagittal and lambdoidal sutures intersect.

m *metopion* is the cranial midline point on the frontal bone where the elevation of the curve is greatest.

ml *mentale* is the most inferior point on the margin of the mental foramen.

n *nasion* is the point in the midline where the nasal bones and the frontal bone intersect.

ns *nasospinale* is the midline point of a tangent between the most inferior points of the nasal aperture.

op *opisthocranion* is the midline cranial point on the occipital bone that is most distant from the glabella.

or *orbitale* is the most inferior point on the orbital margin.

pg *pogonion* is the most anterior midline point on the chin.

po *porion* is the uppermost point on the margin of the external auditory meatus.

pr *prosthion* is the most anterior midline point on the alveolar process of the maxilla.

rhi *rhinion* is the midline point at the inferior end of the internasal suture.

v *vertex* is the highest midline cranial point on the skull, in the Frankfurt Plane.

zm *zygomaxillare* is the most inferior point on the zygomaticomaxillary suture.

zy *zygion* is the most lateral point on the lateral surface of the zygomatic arch.

the various techniques for the dental assessment of age (Schour & Massler, 1944; Hunt & Gleiser, 1955; Hillson, 1996; Whittaker, 2000). Age estimation from dental assessment in juvenile and young adult skeletons is likely to be fairly accurate and is discussed in Chapter 8. Knowledge of dental nomenclature is useful when assessing a skull, and Fig. 3.6 illustrates common dental terms.

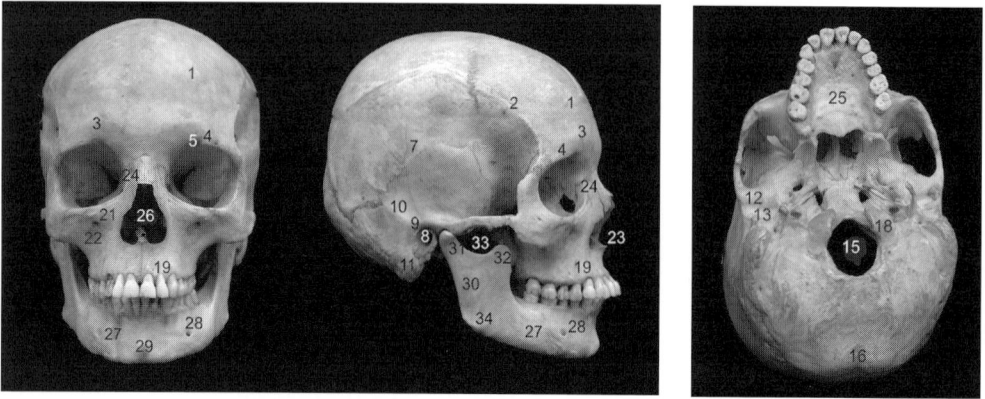

Figs. 3.4 and 3.5 Common cranial landmarks

1 *frontal eminences* – paired frontal bossing.
2 *temporal lines* – mark the attachment of the temporalis muscles on the lateral cranial surfaces of the frontal and parietal bones.
3 *superciliary arches/brow ridges* – bony prominences above the orbits.
4 *supraorbital margins* – the upper orbital edges.
5 *supraorbital notches* or *foramina* – pierce the supraorbital margin to transmit the supraorbital nerve.
6 *metopic suture* – vertical suture between the right and left halves of the frontal bone. Usually obliterates by adulthood.
7 *parietal striae* – striations or rays that pass posterosuperiorly on the cranial surface of the parietal bone from its bevelled squamosal edge.
8 *external auditory meatus* – external opening of the ear canal.
9 *suprameatal crest* – runs horizontally above the external auditory meatus.
10 *supramastoid crest* – raised edge that marks the inferior limit of the temporalis attachment.
11 *mastoid process* – lump on the temporal bone where the sternocleidomastoid, splenius capitis and longissimus capitis muscles attach.
12 *articular eminence* – anterior portion of the temperomandibular articular surface.
13 *mandibular fossa (glenoid fossa)* – lies posterosuperior to the anterior eminence.
14 *styloid process* – thin, pointed, bony rod from the base of the temporal bone.
15 *foramen magnum* – large hole in the occipital bone through which the brain stem passes into the vertebral canal.
16 *external occipital protruberance* – lies on the midline where the occipital and nuchal planes meet.

17 *external occipital crest* – highly variable median line that passes between left and right nuchal musculature.
18 *occipital condyles* – raised oval structures on either side of the foramen magnum.
19 *alveolar process* – horizontal portion of the maxilla that holds the tooth roots.
20 *alveoli* – holes for the tooth roots.
21 *infraorbital foramen* – located below the infraorbital margin and transmits the infraorbital nerve.
22 *canine fossa* – hollow of variable depth located on the facial surface below the infraorbital foramen.
23 *anterior nasal spine* – thin projection of bone on the midline of the inferior surface of the nasal aperture.
24 *anterior lacrimal crest* – vertical crest located on the lateral aspect of the nasal bones.
25 *incisive foramen* – perforates the anterior hard palate at the midline.
26 *vomer* – small, thin, midline bone that divides the nasal cavity.
27 *corpus of the mandible* – the body of the mandible that anchors the teeth.
28 *mental foramen* – large foramen on the lateral corpus surface below the premolar region of the mandible, to transmit the mental vessels.
29 *mental eminence* – triangular bony chin at the base of the corpus.
30 *ramus of the mandible* – vertical part of the mandible that articulates with the cranial base.
31 *mandibular condyle* – large, rounded, articular prominence on end of ramus, which articulates the temperomandibular joint.
32 *coronoid process* – thin, triangular part of ramus where the temporalis muscle inserts.
33 *mandibular notch* – depression between the condyle and the coronoid process.
34 *gonial angle* – rounded posteroinferior corner of the mandible.

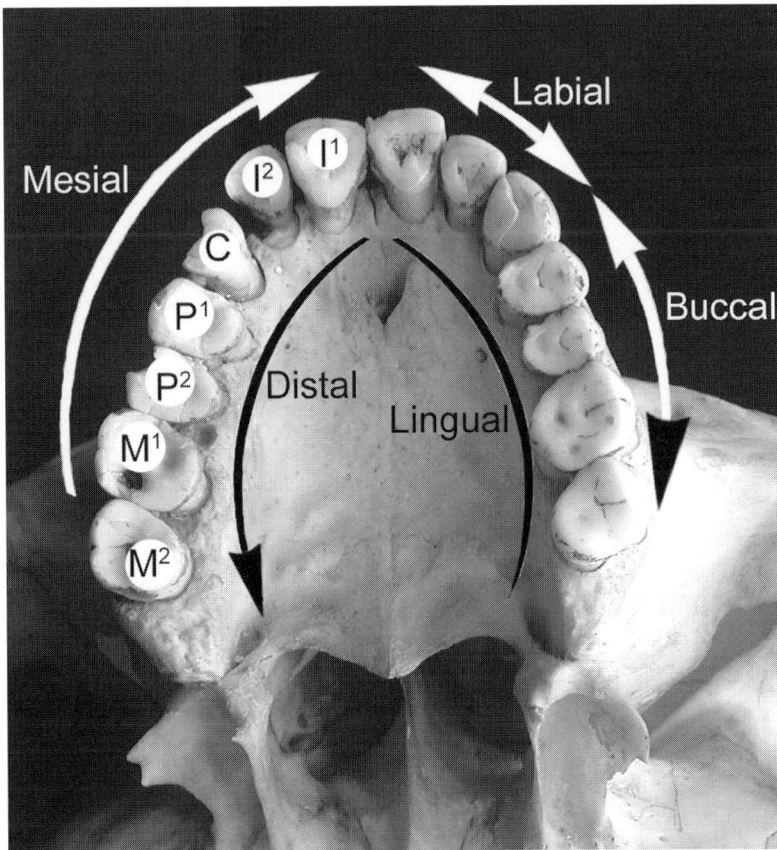

Fig. 3.6 Common dental terms. I^1 = first incisor, I^2 = second incisor, C = canine, P^1 = first premolar, P^2 = second premolar, M^1 = first molar, M^2 = second molar.

Mesial – closest to the point where the central incisors contact each other.

Distal – opposite of mesial.

Lingual – part of the tooth nearest the tongue.

Labial – opposite of lingual, for incisors and canines.

Buccal – opposite of lingual, for premolars and molars.

In addition, the following terms may be useful.

Molars (M^1 and M^2) – two sets of three large tricuspid teeth at back of mouth in maxilla and mandible.

Deciduous (primary) teeth – the first set of teeth to form, erupt and shed in childhood.

Permanent (secondary) teeth – these replace the deciduous teeth.

Crown – the part of the tooth covered by enamel.

Root – anchors the tooth in the bone.

Neck – the junction of the crown and the root.

Enamel – hard tissue covering the crown.

Incisors (I^1 and I^2) – the four spatulate teeth at the front of the maxilla and mandible.

Canines (C) – sharp, pointed third tooth from the midline on maxilla and mandible.

Premolars (P^1 and P^2) – two pairs of bicuspid teeth distal to the canines on maxilla and mandible.

Dentine – tissue forming the core of the tooth.

Pulp chamber – expanded part of the pulp cavity at the crown end of the tooth.

Root canal – narrow end of the pulp cavity at the root end of the tooth.

Cementum – tissue covering the external surface of the roots.

Calculus – calicified deposit on the crowns.

Pulp – soft tissue within the pulp chamber.

Cusp – occlusal projection of the crown.

If at all possible, it is preferable to use microscopic dental section analysis for aging adults. This is considered to be the most accurate and reliable method (Whittaker, 2000). All dental hard tissues show incremental lines of growth and mineralisation. Age-related histological changes in teeth were first suggested by Gustafson (1950), who observed changes in attrition, periodontosis, secondary dentine, cementum, root resorption and root transparency. He created a regression formula to calculate the age of an individual. However, other researchers (Bang & Ramm, 1970) tested his method and found that only root transparency was reliable to use alone. Similar studies from Kilian and Vlček (1989) suggested that more than 75 per cent of their results were within five-year accuracy and, unlike other procedures, an acceptable assessment was possible from a single tooth. Wei *et al.* (1983) studied sections of the first maxillary molar and measured the pulp chamber and dentine to create a pulp–dentine index to determine age. However, estimates may be affected by dental hygiene, occlusion, abrasion, malocclusion, missing teeth, caries, fillings and method of extraction. Incremental lines in enamel can provide an accurate method of age determination. The visualisation of the neonatal line (a pronounced line formed at birth) on deciduous teeth has been described by Rushton (1933), Schour (1936), and Whittaker and Richards (1978), and the recognition of cross-striation patterns was described by Dean and Beynon (1991), and Fitzgerald (1999). The determination of age from microscopic dental assessment appears reliable and accurate, but considerable training and experience are necessary and this has become a specialised field for forensic odontologists and dental experts.

When dental aging by section analysis cannot be performed, the age of an adult skull can be estimated using non-invasive dental aging techniques, which assess dental attrition and calcification (Hunt & Gleiser, 1955). Attrition is the gradual wear of the tooth during mastication by the abrasion of one tooth against another. The pattern of attrition is dependent upon the occlusion of the teeth, and the position of the attrition will indicate how the teeth occlude. Assessment of the attrition pattern can be very useful

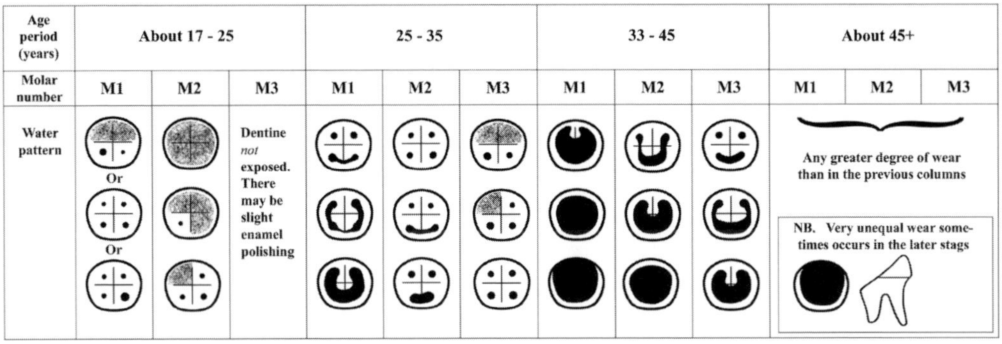

Age period (years)	About 17 - 25			25 - 35			33 - 45			About 45+		
Molar number	M1	M2	M3	M1	M2	M3	M1	M2	M3	M1	M2	M3
Water pattern			Dentine *not* exposed. There may be slight enamel polishing							Any greater degree of wear than in the previous columns NB. Very unequal wear sometimes occurs in the later stags		

Fig. 3.7 Age-related dental wear. Modified from Brothwell (1981). Used by permission of the publishers, British Museum Cornell University Press.

when attempting to set the position of the mandible or when re-assembling a fragmented skull. The abrasive nature of the material in the diet of an individual will have an effect upon the degree of attrition. For example, Ancient Egyptians had a great deal of sand in their diet, due to the desert environment, and that, in combination with poor dental hygiene, led to a substantial degree of dental attrition. In addition, attrition can be affected by the strength of mastication, which may be determined by the fibrous nature of the food, but also by cultural practices. There is evidence, for example, that hide-chewing Inuit women (Pedersen, 1949) and the kat-chewing Yemeni population (Evison, 2002) have abnormal dental wear. It is relatively clear that dental wear becomes gradually more marked as age increases, and some researchers have produced tables of attrition for use with archaeological specimens (Zuhrt, 1955; Miles, 1963; and Brothwell, 1989) (see Fig. 3.7). We can only expect this assessment to be accurate to within ten years, since there is great interpopulation and intrapopulation variation in the wear of teeth and it can be affected by diet, jaw size, sex, trauma and chewing stresses (Brothwell, 1989). This method of assessment is not recommended for individuals over the age of 60 years. Lovejoy (1985) stated that 'assignment of age on the basis of dental wear alone would allow only a gross approximation at best'.

When dental assessment is impossible because of absent teeth, bone assessment will give age estimation to within 10–15 years.

Suture-closure assessment in combination with assessment of the bone texture and muscle attachment strength can give a rough adult-age estimation and tables of suture closure can be found in studies by Krogman and Iscan (1986), Redfield (1970) and Todd and Lyon (1924). Where the bones of the skull meet, a thin membrane may persist unossified for some years after adulthood – this is known as a suture. The bone margins of a suture vary in form and may be flat, saw-edged, overlapping or bevelled (Brothwell, 1989). Normally the sutures begin to close at approximately 20 years of age, and they may eventually become obliterated. Suture-closure age estimation was considered reliable for many years (Parsons & Boc, 1905; Todd & Lyon, 1925), but more recently there was a great deal of criticism (Cobb, 1955; Brothwell, 1981) of the accuracy of such assessments. Cranial suture closure is subject to extreme, inherent variability, and many examples of complete sutural obliteration have been seen as early as 25 years, whilst unfused sutures have been seen in the elderly (Iscan & Helmer, 1993). There may be differences between the sexes and between populations, and some anthropologists (Mann *et al.*, 1991) believe that this assessment can only categorise individuals into '*child, adolescent, young, middle-aged or old adult*'.

The growth of the skull from infancy to adulthood is extreme, and is discussed in Chapter 8. The skull continues to grow throughout adulthood (Behrents, 1985), but these changes appear to be less noticeable on the face than external age-related variations. Changes such as loss of skin elasticity, formation of wrinkles, sagging of flesh, loss of adipose tissue, blurring of the iris detail, increased prominence of facial lines and hair loss can be more indicative of adult age than face size and shape. The skin on the face will become more saggy and droopy, owing to biochemical changes in the connective tissue of the skin that cause it to become less firmly attached to the underlying bone or muscles. Wrinkles appear, due to changes in the distribution and formation of collagenous material in the skin and a decrease in the resilience of the fibres. A decline in the number of fibroblasts in the skin leads to a decrease in secretion and a dehydration of the skin, which also

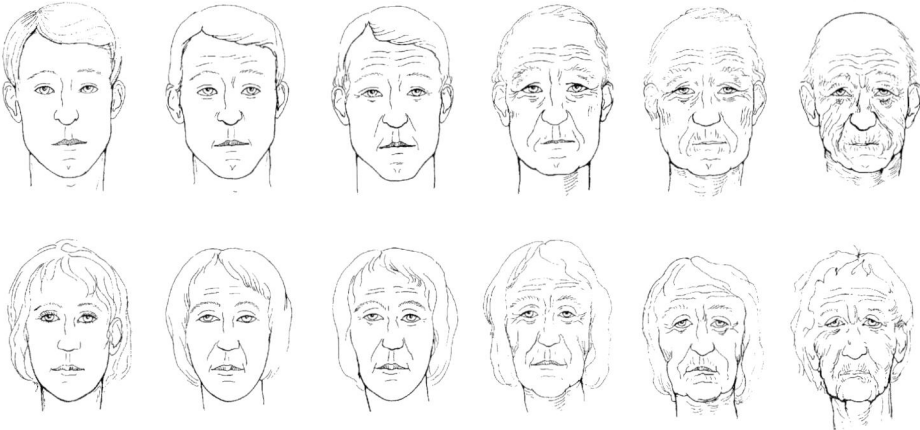

Fig. 3.8 Age-related changes in the adult face. Male (top) and female (bottom). Modified from Neave (1998).

contribute to wrinkles. An old person may appear to have sunken eyes due to resorption of adipose tissue at the orbits and more visible veins beneath the thinner orbital skin, producing dark circles below the eyes. The suborbital region may also begin to sag producing 'bags'. Nasolabial and mental creases will become more marked and deeper with increased age. These age-related changes may be accelerated by other external factors such as cigarette smoking, chronic alcohol consumption, sun damage, medication or loss of weight. A brachycephalic face will appear younger as it resembles the face shape of a child and a dolichocephalic face will appear more mature. A fat face will appear more youthful due to the smoothing of wrinkles and the resemblance to an infant face. Bone resorption at the alveolar processes with loss of teeth in later life will alter the jawline and mouth significantly. The nose and chin will appear more prominent and the distance between the nose and the chin will decrease, with the mouth appearing to sink into the face (see Fig. 3.8). Neave (1998) stated that there is marginal growth of the cartilaginous portions of the nose and the ears in adulthood, and Gerasimov (1971) further stated that there may be some age-related changes in head position. A young child will keep the head directed slightly upwards, whereas at 13–14 years the head position will be near normal (i.e. on the Frankfurt Plane). After 17–18 years the head position drops slowly and a full-grown man has a head position at 12–15° lower than the Frankfurt Plane. To confuse matters, smaller people will keep their heads straighter, with the face directed upwards, and taller people will keep the face directed more downwards.

Sex determination

Sex determination from the skull alone can be problematic. There is inevitably a certain amount of overlapping of the two sexes. Age, environment, pathological changes and, above all, interpopulation variation can influence the determination. Morphological and morphometric traits can both be applied to determine sex and both methods rely on experience. Morphometric measurements may be misleading since total size can be misclassified very easily and large females can be classed as male, and small males as female. Krogman and Iscan (1986) sexed a sample of 750 specimens and found that with the whole skeleton he scored 100%, pelvis alone 95%, skull alone 92% and long bones alone 80%. However, Krogman admitted that these results were somewhat biased since there was a 15-to-1 chance of each individual being male, so his odds of being correct if he said male were quite high. Stewart (1948, 1979) stated that he could correctly sex an entire adult skeleton with 90–95% accuracy and 80% for the skull alone. Stewart also mentioned that Hrdlička hit 80% for crania alone, but if the mandibles were present then that increased to 90%. Stewart stated that, dimensionally, the male-to-female ratio is 100 to 92, i.e. female measurements are 92% of the male measurements. Some studies (Krogman & Iscan, 1986) suggest that the dividing line between immaturity and maturity is 15–18 years.

Morphologically there are many traits assessed in order to sex a skull, and I refer to the work of Krogman and Iscan (1986), White and Folkens (1991), and Iscan and Helmer (1993), who appear to have covered this field in a detailed and thorough manner. The initial

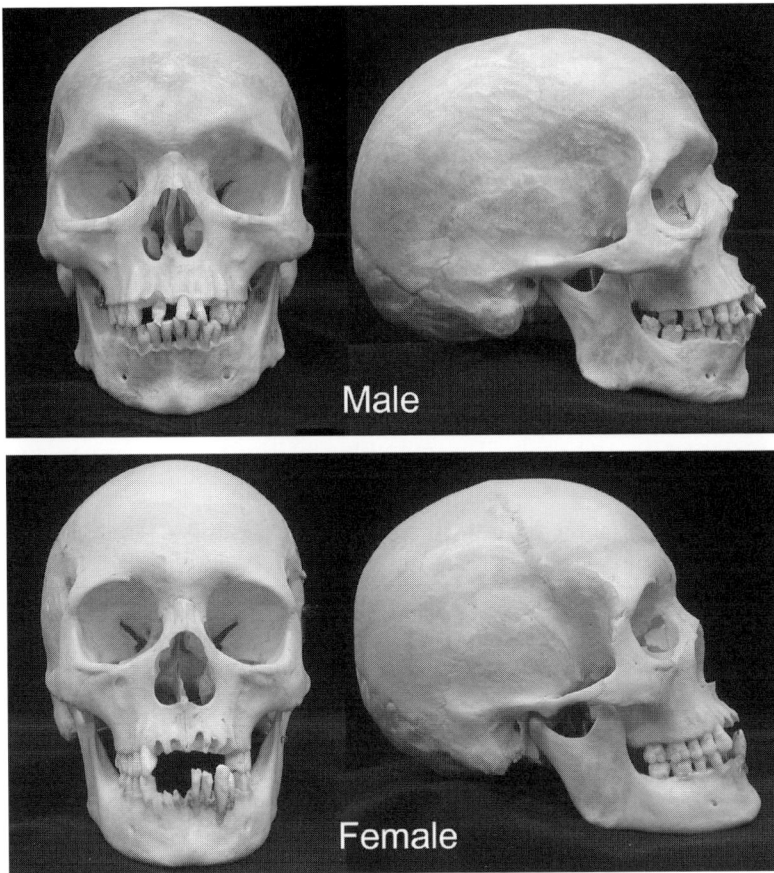

Fig. 3.9 Sexual dimorphism of the skull.

impression of the skull is often the most important assessment.
A large skull will generally belong to a male and a small skull to
a female. The female skull is generally rounder and more gracile
than the male (see Fig. 3.9). These traits are most apparent between
the ages of 20 and 55 years. Different racial origin groups show
different traits and these facts must be taken into account when
assessing a skull (Iscan & Helmer, 1993).

As Fig. 3.9 shows, the forehead contour in the female is higher,
smoother, more vertical and more rounded than in the male. The
male frontal and parietal bones tend to be less bossed than the
female. The supraorbital ridges are more strongly developed in the

male than the female and the glabellar region is larger in the male. The female orbits are higher, more rounded and larger in comparison with the rest of the facial features. The orbital margins are sharper in the female. Male skulls show typically rectangular-shaped orbits, and the nasal aperture is higher, narrower, and its margins sharper. The male nasal bones are larger and tend to meet the midline at a sharper angle. The cheekbones are heavier in the male, and the palate is larger and broader with the arch tending towards a U-shape, as opposed to the more parabolic shape of the female. The occipital region has more evident transverse lines and a larger external protuberance in the male. The mastoid processes are larger in the male. The mandible of the male is larger and thicker, with a greater body height (especially at the symphysis) and with a broader ramus. The gonial angle is less than 125°, the condyles are larger and the chin is more square. Males tend to have more defined muscle attachments, especially at the mandible.

These skull variations lead to obvious differences between male and female faces. The male face is typically larger, with a larger and more protrusive nose, and more flared nostrils (Enlow, 1982). The male forehead is more sloping with more protrusive glabella and supraorbital ridges, giving the male a more upright profile. Due to the greater protrusion of the nose, the male appears to have more deep-set eyes. The female face appears to have eyes that are closer to the front of the face and more prominent cheekbones. All these differences lead to a female face that appears flatter and more delicate, and a male face that is more irregular, coarser and deeper. A female face is more likely to be brachycephalic and a male face is more likely to be dolichocephalic.

These sex-related dimorphic variations may be caused by variations in male and female adult stature. Some researchers (Enlow, 1996) suggest that the difference in the oxygen requirement between males and females may be related to nasal shape, in that the male is typically bigger and more robust, with larger lungs to provide for more massive muscles and organs. This may lead to a larger airway requirement and thus a larger, more prominent nose with more fleshy, flaring nostrils and an accompanying protrusive brow and sloping forehead. The range of the male nasal profile tends to be from convex to straight, and the range of the female nasal profile tends to be from straight to concave. These differences were illustrated by Bruce *et al.* (1993), who studied the surfaces obtained from a number of different male and female faces using laser scanning techniques. Average male and female faces were created from this data, and these average faces showed a larger and more protuberant nose and brow and a broader chin and jawline in the male face than the female face. This gave the impression of more deep-set eyes and less prominent cheekbones in the males, and a wider, flatter face, smaller nose, more upright forehead and more exophthalmic eyes in the females. This may also be illustrated by the suggestion (Enlow, 1982) that female impersonators are more likely to have brachycephalic faces.

As humans, we are very good at determining the sex of an individual from the face. Research by Bruce *et al.* (1993) suggests that without external clues such as hairstyle, make-up, clothing and facial hair, we are 96 per cent accurate at deciding whether a face is male or female, and usually within five seconds (see Fig. 3.10). However, this only applies to adult faces, and problems associated with the sex determination of children's faces are discussed in Chapter 8. Burton *et al.* (1993) investigated the features that were important in judging the sex of a Caucasian face and concluded that the most important variables for facial sexual dimorphism are eyebrow thickness, the width of the nose at its base, the width of the mouth, eye–eyebrow distance, forehead height, the distance between the inner corners of the eyebrows, the distance between the nose point and the lower corner of the nose, and the length of the cheek (from the midpoint of the cheek to the edge of the nose). Another study by Inoue *et al.* (1995) looked at the sex determination of adult Japanese faces by asking volunteers to assess photographs of facial features with regard to gender. They found that the jawline and mouth were very important in gender determination, with less accuracy shown at the eyes and nose.

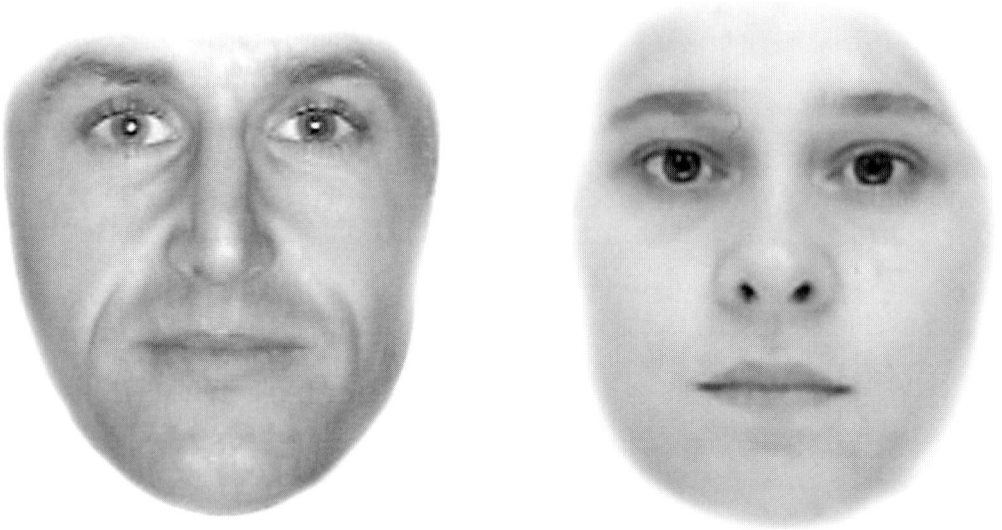

Fig. 3.10 Sexual dimorphism of faces. It is relatively easy to determine which face is male (left) and which is female (right).

Racial origin determination

The determination of racial affiliation is a more difficult assessment. Evolutionary scientists suggest that geographical boundaries separated our Cro-Magnon ancestors early in human history, and biological traits were sustained and perpetuated in isolated gene pools (Landau, 1989). The result of this is that human faces are very varied and linked in geographical terms. This may explain the stereotypical definition of Funk and Wagnall (1961) that, *'race is one of the major zoological subdivisions of mankind, regarded as having a common origin and exhibiting a relatively constant set of physical traits, such as pigmentation, hair form and facial and bodily proportions'*. Some anthropologists (Coon *et al.*, 1950) have tried to explain physical differences as adaptions to the environment. According to this theory the high-arched nose of the Bedouin tribes may be an adaption to humidify the dry desert air before it reaches the lungs; the Inuit's flat, padded face and eye folds may have evolved as protection against the extreme cold; and dark-skinned peoples live in sunny climates where the melanin in their skin protects them from harmful sunlight. Some anthropologists (Davies, 1972; Larsen, 1997) argue that dietary influences can explain the majority of facial types. Larsen states that the Inuit flat-faced appearance is a reflection of masticatory loading rather than cold adaption, and that the zygomas are oriented toward maximising the efficiency and power of chewing. A study of the masseter and temporalis muscle attachments by Spencer and Demes (1993) confirmed this theory. In addition, analysis of cranial dimensions based upon thousands of Native Americans measured by Boas in the late nineteenth century shows no relationship between climate and head shape (Jantz et al., 1992), and many human populations with forward-placed zygomas are associated with hot and dry climates.

However, classifying groups on the basis of facial appearance is fraught with danger. The earth's population has always migrated around the world and genetic mixing may be the rule rather than the exception. The very idea that modern man can be classified into clear-cut subdivisions of racial origin groups is rejected by many anthropologists (Sauer, 1992), osteologists and anatomists alike. Brothwell (1981) states that 'there are skulls which display, say, the Negroid character of strong alveolar prognathism or the flat broad cheeks of a Mongoloid, yet do not belong to either group'. While this is undoubtedly true, it is still very useful in forensic and historical situations to be able to identify the most likely racial origin of an unknown skeleton. It is generally possible to place the remains into one of the major racial origin groups due to certain morphological traits and features. However, there are broad overlaps between racial origin groups and much variation within groups that may be due to the intermingling of races throughout the centuries, leading to a 'blurring' of the margins of traditional racial origin groupings. This may also be a reflection of natural variation and currently such a panmixia exists that subdivisions by appearance can be defined only very broadly (Stewart, 1979). Hooton (1946) states that 'a race is a great division of mankind, the members of which, though individually varying, are characterised as a group by a certain combination of morphological and metrical features, principally non-adaptive, which have been derived from their common descent'.

The racial origin groups most commonly used are Caucasoid (including Europeans, Asians from the Indian subcontinent, North and East Africans, Arabs and Mediterraneans), Negroid (including West and South Africans) and Mongoloid (including Asiatics, Inuits and Native Americans) and Australoids (Australian Aborigines, Pacific Islanders, Fijians and Papuans) (Cole, 1965). There is some dispute over the finer subdivisions of these groupings and Polynesians, Australian Aboriginals and Native Americans have received considerable anthropological investigation and may require racial origin groups of their own. It is thought that Native Americans were derived from Mongoloid origins, which separated from the Asiatic ancestors approximately 10 000–15 000 years ago (Boyd, 1950; Dunn, 1967). Some anthropologists (Cole, 1965; Iscan & Helmer, 1993) further subdivide these basic racial origin groups. It is claimed (Clement & Ranson, 1998) that there is a 77–95 per cent accuracy of race determination from a skull. However, these traits are not clear until puberty and the racial determination of a preadolescent skull is more difficult (see Chapter 8).

The Caucasoid skull has a long-to-rounded shape with a narrow nasal aperture, moderate supraorbital ridges, blunt supraorbital margins, sharp lateral orbital margins, a depressed glabella, beetling of the frontonasal junction, sharp cheekbones, a prominent frontal aspect, large mastoid processes, tortuous cranial sutures, a prominent nasal spine, deep canine fossae, mild or no prognathism, an undulating lower border of the mandible, a steepled nasal root and a narrow interorbital distance. This leads to a steeper forehead with no sagittal plateau, an orthognathic profile, a prominent chin and a variable occipital curve. (Brues, 1990; Clement & Ranson, 1998). Gerasimov (1971) stated that an angular orbit with a square or rhombic outline and blunt angles is typically Caucasoid (see Fig. 3.11). Enlow (1996) stated that Caucasians are more likely to exhibit dolichocephalic head shapes with a more protrusive upper face and more retrusive lower face. This gives a more forward placed nasomaxillary region and a tendency for the downward and backward rotation of the mandible. The palate will

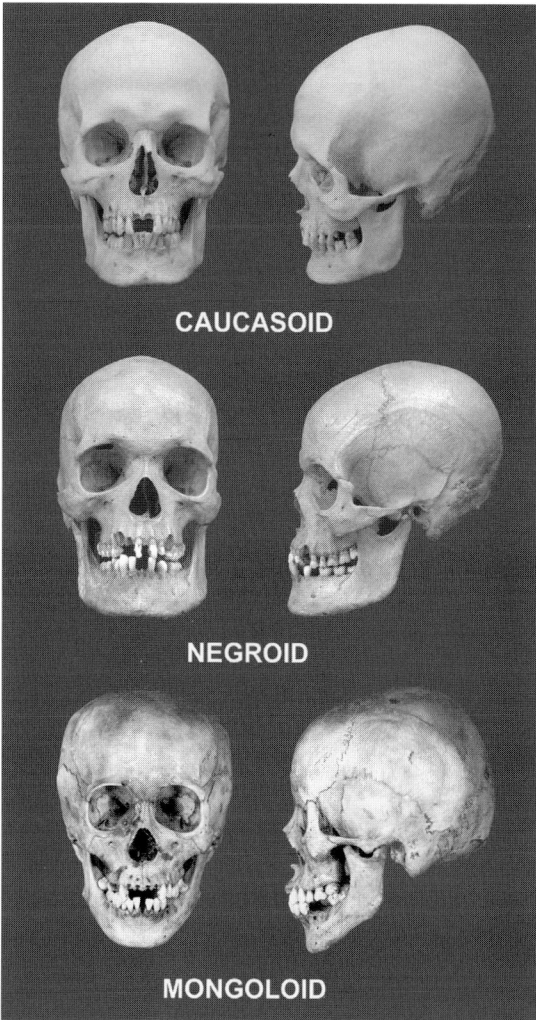

Fig. 3.11 Racial origin variation in skull morphology.

therefore be long and narrow with a large frontal sinus and protrusive glabella. The eyeballs will be deep-set, the cheekbones less
prominent and the profile convex. Enlow claimed that this typical Caucasoid shape leads to a built-in Class I dental occlusion
tendency of maxillary protrusion and/or mandibular retrusion.

The Negroid skull shows a long head shape with wide nasal
aperture, more undulating supraorbital ridge, strong alveolar

prognathism, low rounded nasal root, sharp upper orbital margins, guttering of the inferior piriform aperture, rounded glabella, plain frontonasal junction and wide interorbital distance. This leads to a fuller, more rounded forehead, flatter sagittal contour, postbreg-matic depression, dolichocephalic skull and a full, rounded occip-ital contour (Krogman & Iscan, 1986; Brues, 1990) (see Fig. 3.11). Enlow (1996) stated that Negroid skulls tend to be dolichocephalic, as with most Caucasoids, but with a much less protrusive upper face. This means that the forehead is much more upright and bul-bous than most Caucasians, the frontal sinus is less expanded, the nasal bridge lower, the nose flatter, wider and less protrusive, and the cheekbones more prominent. The mandibular ramus tends to be very broad, which places the mandibular corpus in a resultant protrusive position. This causes the maxillary incisors to tip labi-ally, and this creates bimaxillary protrusion and tends to preclude a Class II type malocclusion. Hajnis *et al.* (1994) studied anthropo-metrical differences between the three racial origin groups (North American Whites, Black Africans and Chinese) and found that Black Africans had dolichocephalic or relatively elongated heads, being the largest in length and circumference and smallest in width. The facial index indicated a long face type, with a long lower face. Black Africans showed the longest interocular distance, longest fissure length and widest intercanthal distance. The Black African nose was wide, flat and short and the mouth was the widest, showing the largest lip and vermilion heights. Hajnis *et al.* also found that Black Africans had the shortest and widest ears.

The Mongoloid skull shows a round head shape with a medium-width nasal aperture, rounded orbital margins, massive cheek-bones, weak or absent canine fossae, moderate prognathism, absent brow ridges, simple cranial sutures, prominent zygomatic bones, broad, flat, tented nasal root, short nasal spine, shovel-shaped upper incisor teeth (scooped out behind), straight nasal profile, moderately wide palate shape, arched sagittal contour, wide facial breadth and a flatter face (Brues, 1990; Clement & Ranson, 1998) (see Fig. 3.11). Enlow (1996) stated that Mongoloid skulls are more likely to be brachycephalic, with foreshortened palates and dental arches. The midface is less protrusive with a wider, shorter nasal airway. The forehead is more upright and bulbous, has a less protrusive glabella and supraorbital ridges, a lower nasal bridge, a shorter pug-type nose, shallow orbits, less deep-set eyeballs and a tendency for forward rotation of the mandible. The face appears flatter, broader and squarer with prominent cheekbones and a straighter profile. There is a greater tendency for Class III malocclusion, and a prognathic mandible exists. Hajnis *et al.* (1994) studied anthropometrical differences between the three racial origin groups and found that Chinese heads are the shortest and most brachycephalic, being the largest in width and smallest in length and circumference. The facial index suggests a mesocephalic-type face, with the largest intercanthal width and shortest eye fissure length. The Chinese nose was found to be mesorrhine or medial, protruding less and being wider than the Caucasian nose, but protruding more and being narrower than the Black African nose. The results also showed that the Chinese faces had the narrowest mouth width and smallest upper lip height, and exhibited the narrowest ears.

The Aboriginal skull shows a flattened sagittal contour of the vault with a broad nasal aperture, low nasal bones, no external occipital protuberance, distinct occipital torus, moderate to large supraorbital arches, moderate to large glabella, medium zygomatic bones, dolichocephalic shape and marked subnasal prognathism (Lanach, 1978; Pounder, 1984). These traits will give the Aboriginal face a low, flat nose, pronounced brow ridges and a prognathic profile.

The Polynesian skull has a high vault, prominent parietal bossing, absent or reduced nasal bones, zygomatic bones that are visible from above, orthognathic profile, large upper facial height, minimal nasal prognathism, prominent chin, curved inferior border of the mandible (rocker jaw) and high mandibular rami (Snow, 1974; Houghton, 1978). These traits will give the Polynesian face a low, flat nasal profile, wide flat cheekbones, an upright profile, rounded jawline and long lower face.

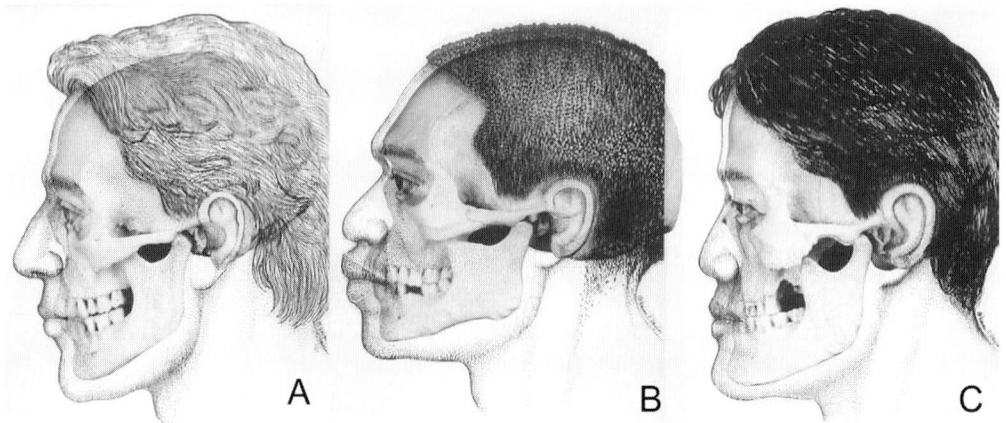

Fig. 3.12 Facial differences related to racial origin. A = Caucasoid, B = Negroid, C = Mongoloid. Courtesy of Karen Taylor (2001).

Schultz (1918) showed racial differences at the mandible between Blacks and Whites. He stated that the Caucasoid skull showed a larger mandible with a higher ramus, greater gonial angle, strong masseter attachment and more protrusive chin; whilst the Negroid skull showed a lower, wider mandible with a more vertical ramus, greater dental arch length and a less dominant chin. However, similar studies (Morant, 1936) showed that racial differences in the mandible were virtually non-existent, and dimensional variation in the mandible was so large that any racial differences were not significant (Jankowsky, 1930). Some forensic anthropologists have studied racial differences by metrical means and the works of Giles and Elliot (1962), and Howells (1970) are extensive and suggest an 80–88 per cent accuracy. Stewart (1948) stated that *'the chances of being right in a racial identification depends largely upon the observer's experience'*. Some researchers (Johnson et al., 1990) have developed methods of race determination from linear and angular dimensions. People today rarely fall into such convenient categories, but general classifications can be helpful in a forensic investigation (see Fig. 3.12) to focus on the population from which an individual may be identified.

The differences in skull morphology between the racial origin groups will inevitably lead to differences in facial morphology between those groups and there has been a great deal of research into these anthropometrical differences. The roots of this research lie in the unpalatable field of physiognomy and craniometrics, made infamous by della Porta, Broca and Galton (see Chapter 1, p. 8), whereby facial morphology was used to determine temperament, personality and superiority. However, many contemporary studies into the facial differences between ethnic groups have been carried out within the fields of maxillofacial surgery, dentistry and physical anthropology. Farkas *et al.* (2000) studied the frequency of seven neoclassical facial canons in African-American and North American Caucasian populations and found that the three sections of the facial profile were not equal in the two populations. They found that African-Americans had longer lower faces and wider noses than North American Caucasians. A similar study of Chinese and North American Caucasians (Wang *et al.*, 1997) found that Chinese adults exhibited nasal widths equal to (52%) or wider than (26%) one-quarter of the face width, whereas North American Caucasians exhibited nasal widths that were equal to (37%) or narrower than (39%) one-quarter of the face width. They also showed that the mouths of Chinese people were more frequently (72%) narrower than 1.5 times the nasal width, while the North American Caucasian mouths were more frequently (60%) wider. Some studies (Ofodile *et al.*, 1993) suggest that Black noses more frequently (53%) exhibit concave dorsal ridges and are shorter and wider than White noses.

Plastic surgery studies typically split noses into three basic ethnic variations: the leptorrhine for Whites, platyrrhine for Blacks and mesorrhine for Asians. The platyrrhine nose is typically the widest (40 mm), with the leptorrhine being the narrowest (31.4 mm) and the mesorrhine in the middle (35.5 mm). One study (Milgrim *et al.*, 1996) found that Latino noses should also be classed as mesorrhine, and another (Sanchez, 1980) coined the type 'chata' to describe the Puerto Rican nose as having a short broad dorsum, wide flat tip, wide alar base, overhanging ala and thick skin. Hajnis *et al.* (1994) studied anthropometrical differences between the three racial origin groups by measuring the faces of 103 North American

Whites, 100 Black Africans and 60 Chinese. They found that
the Chinese showed brachycephalic heads, Black Africans showed
dolichocephalic heads and Caucasians showed mesocephalic heads.
Their results suggested that the Black African head was largest in
length and circumference and was narrowest, whilst the Chinese
heads were widest and smallest in length and circumference. The
facial measurements suggested that the Caucasian males and both
the male and female Chinese were defined as the medial face, and
the Caucasian female and both sexes of Black Africans were defined
as long-faced. The Chinese face was widest, the Caucasian face had
the smallest vertical profile, and the Black African face had long
vertical proportions but was relatively narrow. The Chinese group
showed the largest intercanthal width and the shortest eye fissure
length, Black Africans showed the largest interocular distance, and
Caucasians showed the largest eye fissure height. The Caucasian
nose was defined as leptorrhine or narrow and protruding, the
Chinese nose as mesorrhine or medial and the Black African nose
as platyrrhine or wide and flat. Black Africans exhibited the widest
mouths and the largest lip heights, the Chinese group exhibited
the narrowest mouths and the smallest upper lip height, and the
Caucasians exhibited smallest vermilion height. The ears of the
Caucasian group were the longest, and the shortest and widest
were seen in the Black African group. The Chinese group showed
long and narrow ears. Other studies (Sim & Smith, 2000; Dibbets &
Nolte, 2001) agree with these results. Another study of the soft
tissue profiles of Koreans and White Americans (Hwang *et al.*, 2001)
showed that Koreans had more prominent lips, and less prominent
chins and noses than White Americans. This study analysed 60
Koreans and 42 White Americans from profile tracings. Another
study (Hanihara, 2000) compared the frontal facial features of 112
populations from around the world, and found that very flat faces
were seen in East Asian faces. They also found that deep nasions,
marked prognathism and flat frontal bones were characteristic of
Australian/Melanesians, and that subSaharan Africans shared the
marked prognathism and flat frontal bones, but also showed flat
nasal and zygomaxillary regions similar to East Asians.

Other factors determined from the skull

Some other less predictable factors may be determined from the skull, such as pathological or possibly even cultural indications. Many diseases may be apparent from the cranial remains (Aufderheide & Rodriguez-Martin, 1998; Cox & Mays, 2000) and afflictions such as leprosy, osteomyelitis, tumours and dental abscess formation may be diagnosed from the skull, whilst certain genetic dispositions, such as Treacher–Collins syndrome, cleft palate and hydrocephaly may also be determined. For example, Robert the Bruce, the Scottish battle hero, is thought to have suffered from long-standing leprosy, which could be determined from the loss of maxillary alveolar bone, maxillary teeth and nasal bone (see Fig. 3.13). The facial reconstruction of Robert the Bruce, by Richard Neave,

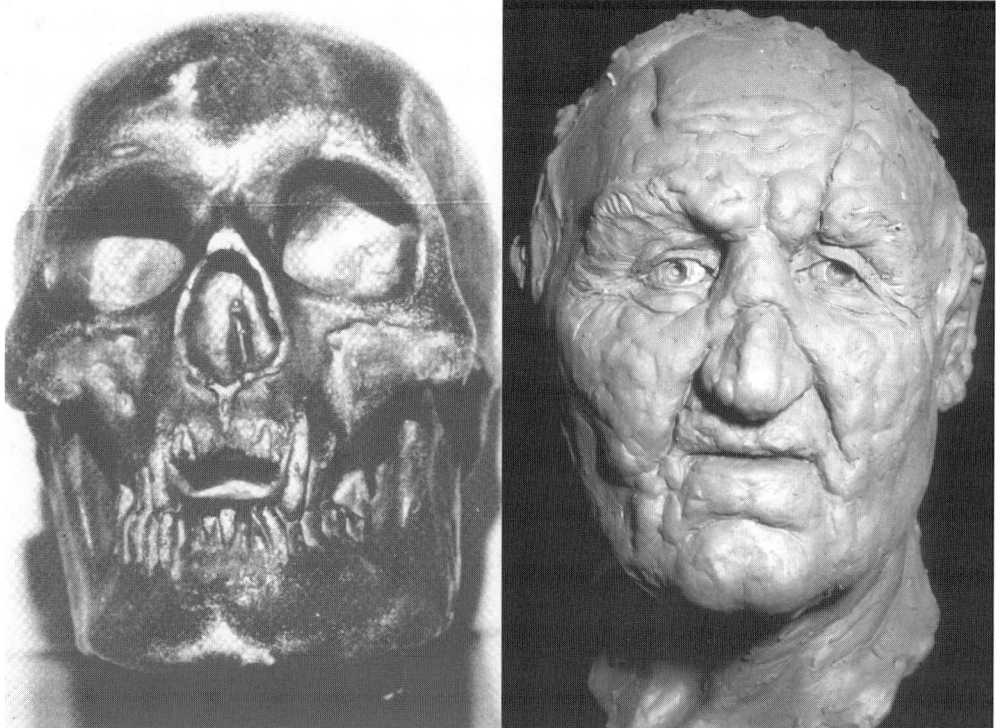

Fig. 3.13 The effects of long-standing leprosy on the skull and face. The skull and facial reconstruction of Robert the Bruce by Richard Neave.

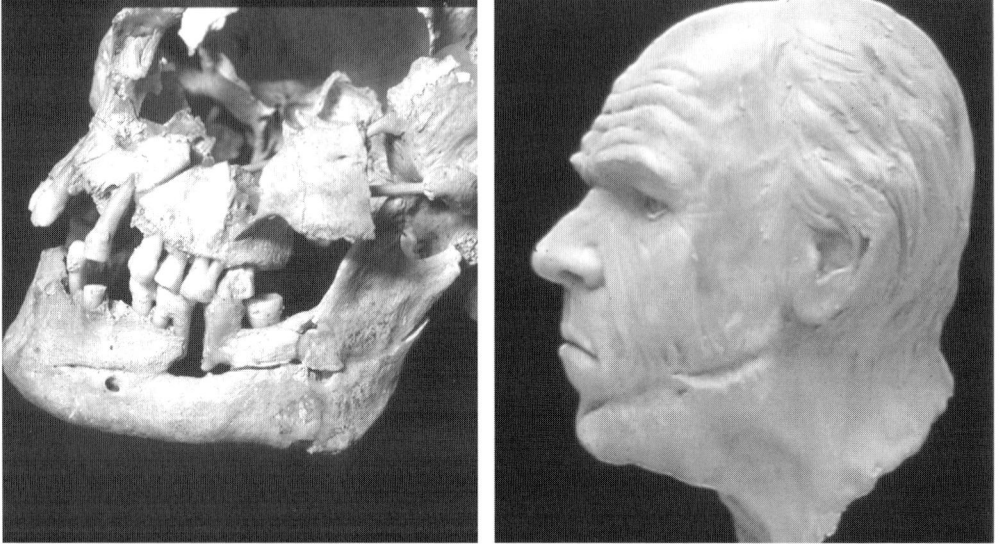

Fig. 3.14 Skull and reconstruction of a soldier from the Battle of Towton (1471). The skull exhibited a healed wound to the mandible from a heavy blow to the side of the head with a sword-like implement. The wound was at least ten years old and the facial wound would have been extensive. The soldier would have received some medical treatment but a rough scar may have remained.

included the common facial appearance associated with leprosy, such as the saddle-back nose and the pitted, granular surface to the skin.

In some cases there may be healed wounds that can be assessed from the skeletal structure of the skull (Fiorato *et al.*, 2000). For example, Fig. 3.14 shows the skull of a 50-year-old soldier from the Battle of Towton (Wilkinson & Neave, 2003). The soldier exhibited a healed wound to the mandible, with bone reformation that suggested the wound was at least ten years old. The wound was deep and long, and had penetrated through the soft tissue to the palate and almost severed the mandible. This wound was consistent with a sword blow and would have severed a facial nerve and led to partial facial paralysis.

With respect to cultural factors, determination from skull remains would be limited and rare but, on occasion, unusual bone deformation may indicate cultural influences. For example, it was common practice in ancient Turkey (900 BC) to bind the head of

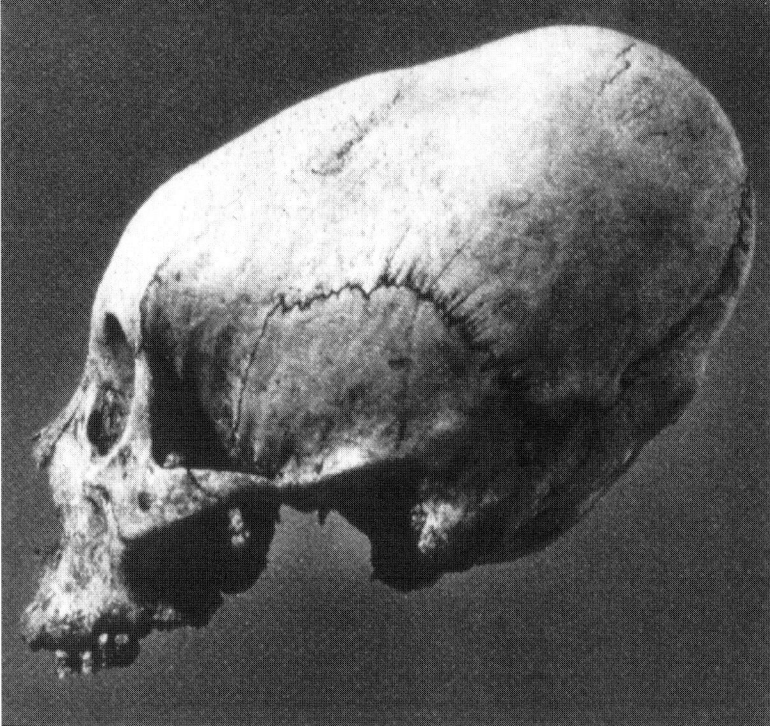

Fig. 3.15 Skull of King Midas. This skull has a long and elongated vault, thought to have been caused by head-binding during infancy. Long dolichocephalic heads were considered attractive in Ancient Turkey (900 BC). (Prag & Neave, 1997).

an infant to create deformation of the skull bones into a narrow, elongated head shape, which was considered fashionable in that society (Prag & Neave, 1997; see Fig. 3.15). Other cultural indications include dental mutilation (filing of teeth to points or to include holes), and skull tattoos. Many other studies have assessed the finer details of the skull and have determined how facial morphology can be established from the skeletal remains.

4 The relationship between hard and soft tissues of the face

Knowledge of common morphological terms is necessary when carrying out an assessment of the face. Figs. 4.1 to 4.5 illustrate these facial surface anatomical terms. There has been a long history of research into quantifying the relationship between the skeletal structure of the skull and the overlying soft tissues of the face, with the express purpose of facilitating facial reconstruction. Gerasimov (1971) was convinced that there was a clear correlation between the relief of the skull and the surface of the soft stratum. He claimed that this could be illustrated by the asymmetry of the skull, which is also exhibited in the asymmetry of the face. He clarified this with a photographic study of faces to create totally right-sided and left-sided faces from an original face. He took a frontal-view photograph of a subject and split the photograph on a line that dissected the glabella and the philtrum. He then created two faces by mirror imaging each side and attaching it to its mirror image: one made up of two right sides and one of two left sides. He found for every subject two distinct faces were created: a 'fine' face and a 'rough' face. He suggested that asymmetry is a basic element of individuality and that since the asymmetry is natural, any reconstruction of the soft tissues will define the character of this asymmetry and secure similarity to the actual face (see Fig. 4.6). Most faces are surprisingly asymmetrical and this seems to be universal among different cultures and nationalities, whilst asymmetry is visible in foetuses as early as 14 days. Yet we are usually unaware of the asymmetry in the faces we view day to day, as our brains tend to accommodate visual imbalances. In addition, facial expression, movement and head position will make asymmetries less apparent, and any asymmetry would have to be quite marked to be noticeable in every day life. Indeed, most of us could be recognised from

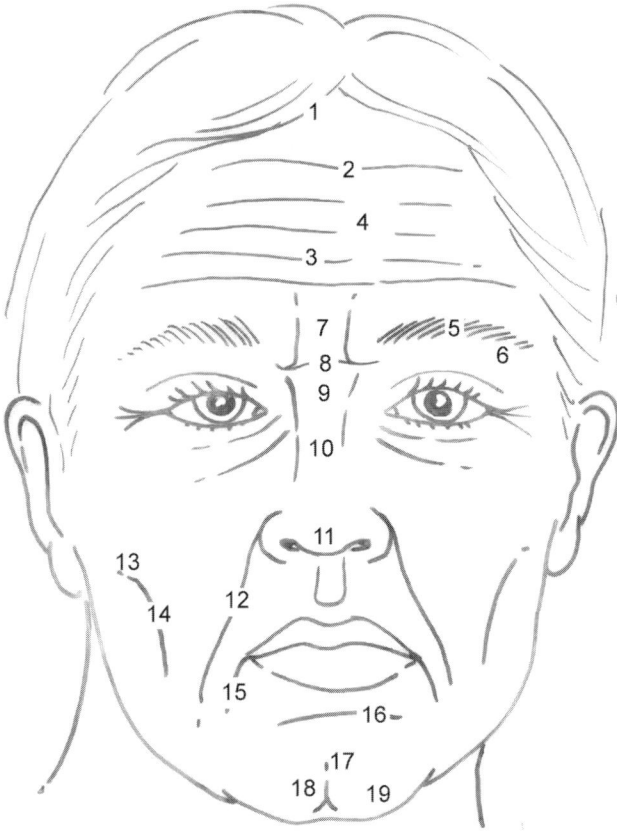

Fig. 4.1 Facial anatomy terms.

Modified from Dunn and Harrison (1997).

1	hairline	11	nasal tip
2	upper forehead crease	12	nasolabial crease
3	forehead creases	13	buccal pit
4	forehead	14	buccomandibular groove
5	eyebrow	15	marionette line
6	supraorbital margin	16	mental crease
7	vertical glabellar lines	17	mental pit
8	transverse nasal groove	18	chin crease
9	nasal root	19	chin
10	nasal bridge		

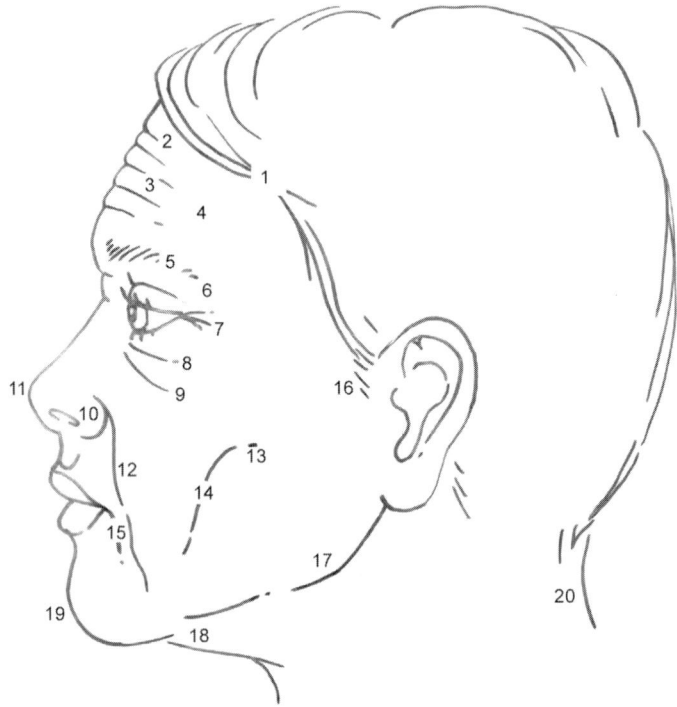

Fig. 4.2 Facial anatomy terms.

Modified from Dunn and Harrison (1997).

1	hairline	11	nasal tip
2	upper forehead crease	12	nasolabial crease
3	forehead creases	13	buccal pit
4	forehead	14	buccomandibular groove
5	eyebrow	15	marionette line
6	supraorbital margin	16	sideburn
7	lateral canthal creases	17	angle of mandible
8	infraorbital crease	18	chin–neck angle
9	lower lid crease	19	chin
10	ala/nasal wing	20	nape of neck

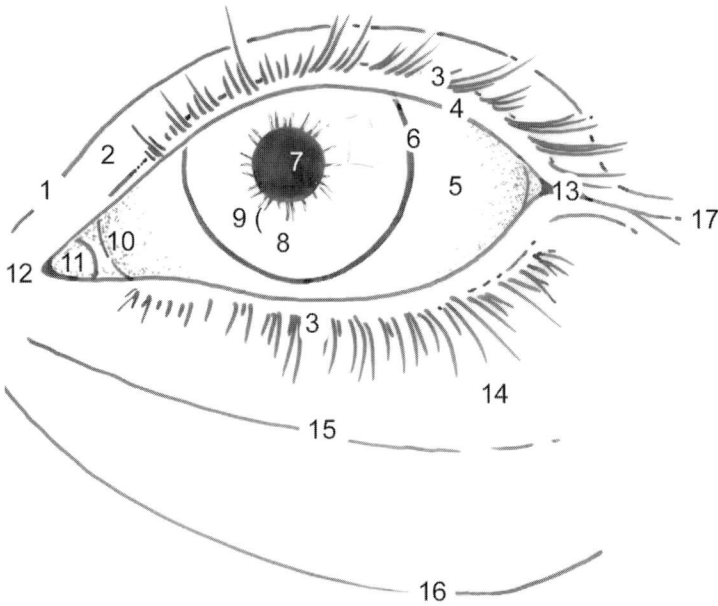

Fig. 4.3 Anatomy of the eye.

Modified from Dunn and Harrison (1997).

1	upper lid crease	10	plica semilunaris
2	upper eyelid	11	lacrimal caruncle
3	lash margin	12	medial canthus
4	grey margin	13	lateral canthus
5	sclera	14	lower lid
6	limbus	15	lower lid crease
7	pupil	16	infraorbital crease
8	iris	17	lateral canthus creases
9	cornea		

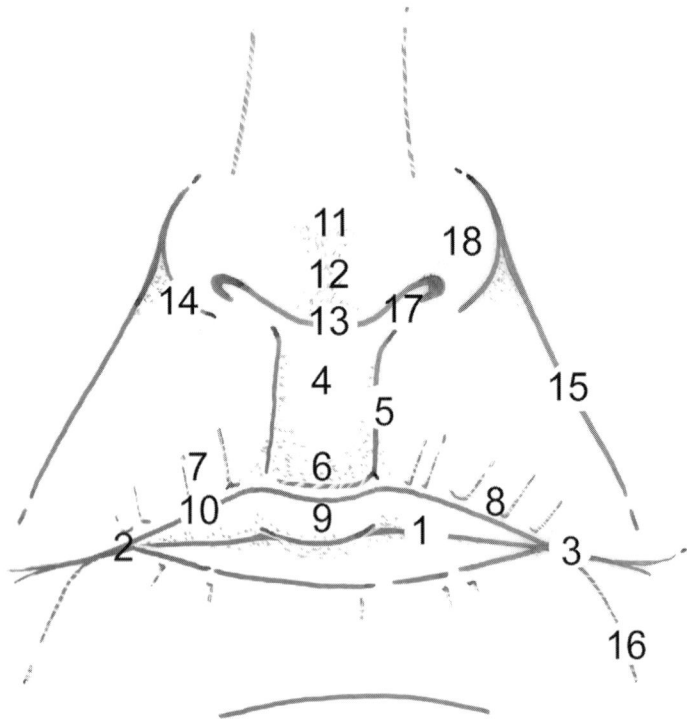

Fig. 4.4 Anatomy of the mouth.

Modified from Dunn and Harrison (1997).

1	oral fissure	10	vermilion border
2	chelion point/corner of mouth	11	nasal tip
3	commissure crease	12	nose tip groove
4	philtrum	13	columella
5	philtral column	14	alar crease
6	cupid's bow	15	nasolabial crease
7	circumoral site/rhytides	16	marionette line
8	white roll	17	nostril
9	lip tubercle	18	ala/wing of nose

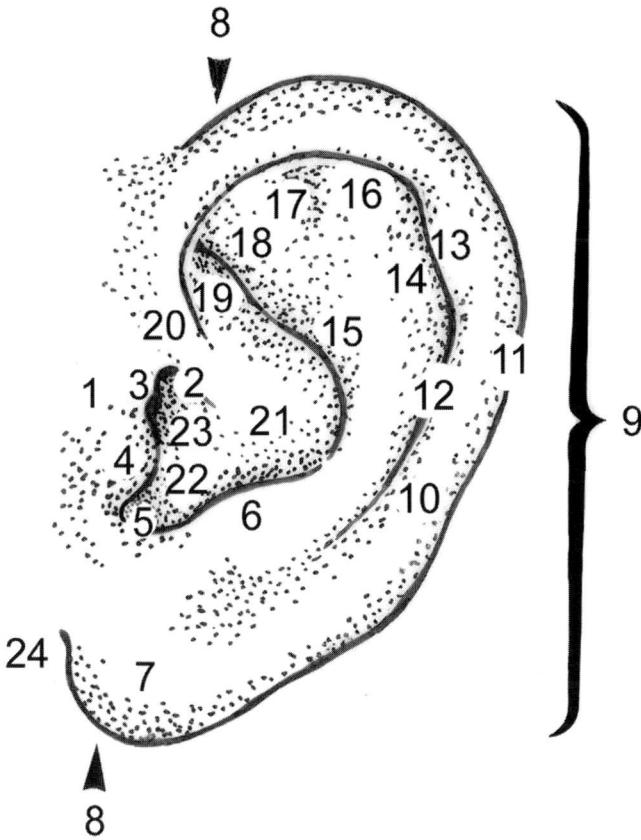

Fig. 4.5 Anatomy of the ear.

Modified from Dunn and Harrison (1997).

1	preauricular sulcus	13	Darwin's tubercle
2	anterior notch	14	scaphoid fossa
3	supratragal tubercle	15	antihelix
4	tragus	16	upper crus of antihelix
5	intertragal notch	17	triangular fossa
6	antitragus	18	lower crus of antihelix
7	lobe	19	cymba concha
8	posterior auricular sulcus	20	crus of helix
9	pinna	21	conchal fossa
10	helix	22	conchal cave
11	helical rim	23	external meatus
12	helical margin	24	terminal notch

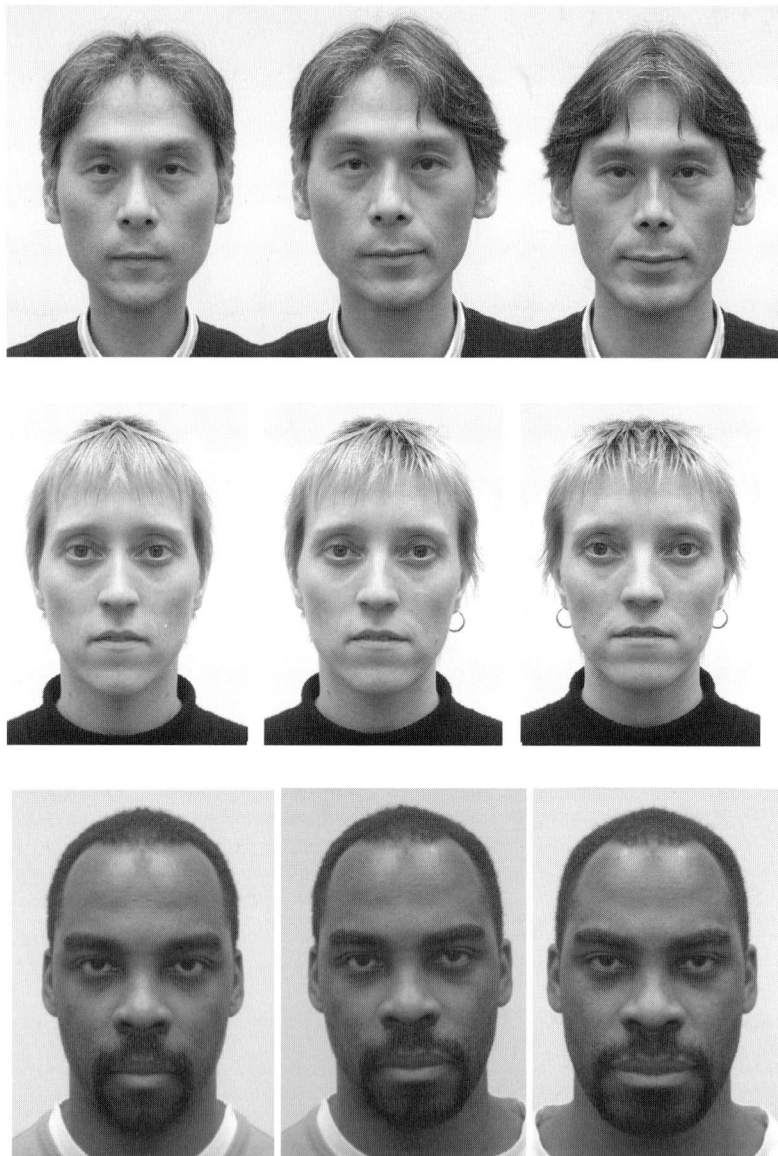

Fig. 4.6 Examples of normal facial asymmetry. Original photograph of the individual (middle). Composite image made up of two left sides (left). Composite image made up of two right sides (right).

a composite image and the asymmetry would have to be very great (see Fig. 4.7) before an individual could not be recognised.

Broca is considered by many (Fedosyutkin & Nainys, 1993) to be the first researcher to study the relationship between the structure

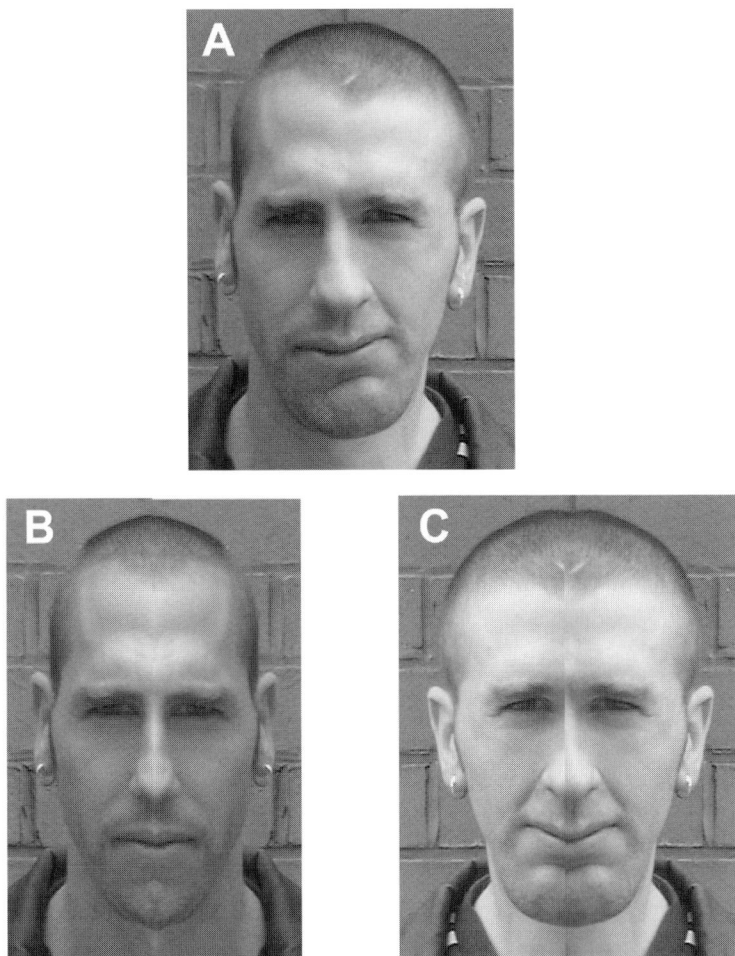

Fig. 4.7 An example of unusual asymmetry in the face. Original photograph of the individual (A). Composite image made up of two left sides (B). Composite image made up of two right sides (C).

of the skull and the overlying soft tissues that define appearance. He noted great individual variation in soft tissue thickness from one person to the next. At the turn of the century criminalists drew the attention of anatomists to the results of research into the correlation between the structure of the soft tissue formation of the nose with the configuration of the piriform aperture. This led to a spate of research into the comparison of faces with skull structure, and many different, often grossly discordant hypotheses arose.

The general face shape

The shape of the upper head is described by Fedosyutkin and Nainys (1993) using four terms: rounded, square, oval and triangular. These shapes are determined by the transverse arc (vault) of the cranium which is described as: semisphere, pentagonoid, oval and rectangular respectively (see Fig. 4.8). The shape of the lower part of the face repeats the contour of the mandible and, if the gonial angle is over 125° and the coronoid process is high, then the lowest part of the face is likely to be a narrow variant, such as oval or

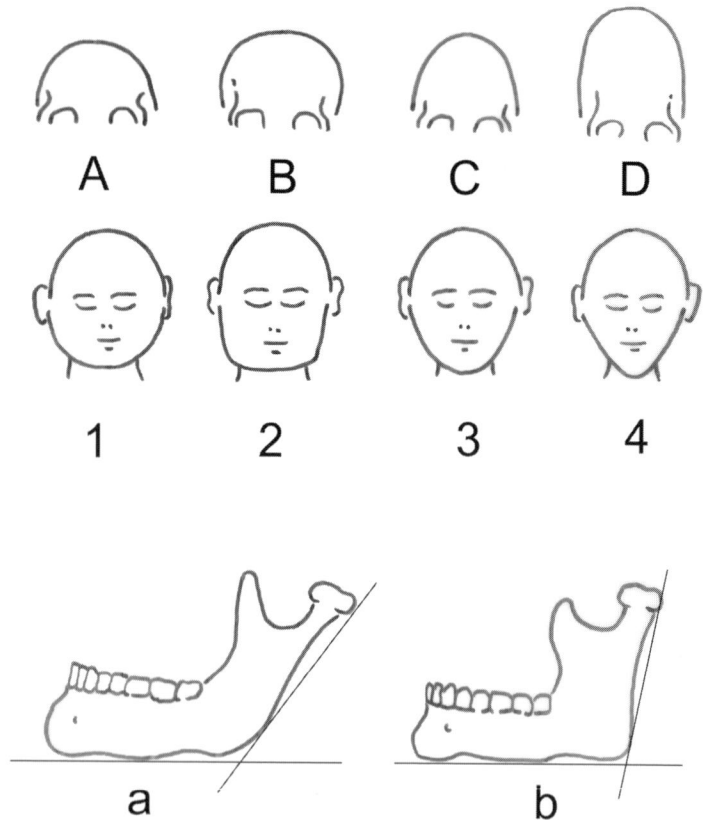

Fig. 4.8 Face shapes. Semisphere (A), pentagonoid (B), oval (C) and rectangular (D) vault shapes. Round (1), square (2), oval (3) and triangular (4) face shapes. Obtuse (a) and acute (b) gonial angles. Modified from Fedosyutkin and Nainys (1993).

triangular. If the gonial angle is less than 125° and there is a wide, low coronoid process, then the face shape is likely to be a wide variant, such as rounded or rectangular. Fedosyutkin and Nainys also claimed that the hairline could be seen microscopically as the transition of the smooth surface of the forehead bone into small rough tubercles. Some studies (Gerasimov, 1971) suggested that when the cheekbones are flat, the zygomatic muscle is located on a more frontal surface of the zygomatic bone; and when the cheekbones are strongly profiled, the zygomatic muscles are located mainly on the lateral surface of the zygomatic bone.

The nose

The majority of early facial anthropological research seemed to centre around the nasal form. Tandler (1909) stressed the importance of the nose in facial reconstruction but did not believe that the configuration of the nasal tissue was correlated to the contour of the bones at the root of the nose. His (1895), Birkner (1907) and Virchow (1912) studied the correlation of the prominence of the nasal spine and nose protrusion and found variations in tissue thickness based on racial origin (see Chapter 5). A study of the relationship between the external nose and the bony nose and nasal cartilages (Schultz, 1918) in 8 white cadavers and 23 black cadavers found that the height of the external nose (distance between the nasal and subnasal points) corresponded to the nasion–subspinal distance on the skull. Schultz also found that in adults the subnasal point lay below the subspinal point (1.4 mm in Whites and 1.6 mm in Blacks), but at birth the subnasal point was found to lie above the subspinal point. This explains the fact that a newborn's nostrils lie above the floor of the nasal cavity, while in the adult they lie below. Schultz found that the breadth of the nose was always greater than the piriform aperture by an average of 10 mm in Whites and 15 mm in Blacks. Other studies of the noses of African Blacks, Chinese, Indian and Sunda Islanders (Virchow, 1912) found similar relationships (13 mm in Blacks, 15 mm in Chinese, 13.5 mm

in Indians and 12.5 mm in Sunda Islanders) between the breadth of
the nose and the nasal aperture. Schultz, however, concluded that
the wide variation in nasal breadth and nasal aperture breadth
meant that there was no standard correlation between the two.
The results of many later studies are in disagreement with this conclusion and are discussed later in this section. Schultz (1918) also
stated that the nose was the most distinctive racial characteristic,
and he found many racial differences between Blacks and Whites in
nasal cartilage form, which seemed to be larger in Whites. Virchow
reinforced these findings with similar studies into the nasal cartilages of Blacks and Japanese. This would concur with the theory
of a more prominent nose in Whites that needs more cartilaginous support and a more protruding nasal septum. The average
length of the nasal septum in Whites was found to be 4.75 mm,
compared to 2.6 mm in Blacks. Glanville (1969) studied the nasal
shape and prognathism in 167 Caucasoid skulls and 165 Negroid
skulls. He found that increasing prognathism was associated with
an increasingly broad and short nose. He suggested that a high correlation existed between the nasal height and the length of the cranial base, and between nasal breadth and the distance between the
upper canine teeth. The soft parts of the nose are a natural continuation of the bony part and in order to accurately assess the shape
of the nose, the glabella, brow, eyes, maxilla and cheek bones must
be assessed. Gerasimov stated that 'an examination of the nasal bones
alone does not suffice for a reconstruction and is really pointless'. His studies found that usually a prominent glabella will accompany prominent brow ridges, and the nasal spine and accompanying nasal
bone define the general form of the profile of the nose and the
degree of prominence of the nose in front (see Fig. 4.9). Gerasimov
also found that the height of the wings of the nose is influenced by
the position of the crista conchalis of the lower muscle in the nasal
opening. Gerasimov suggested that the shape of the lateral nasal
bones and the outline of the nasal aperture will determine the alar
shape. The lateral nasal bones will often change direction sharply
along the outline of the nasal aperture and this point will determine the superior edge of the alar wings. Any asymmetry in the

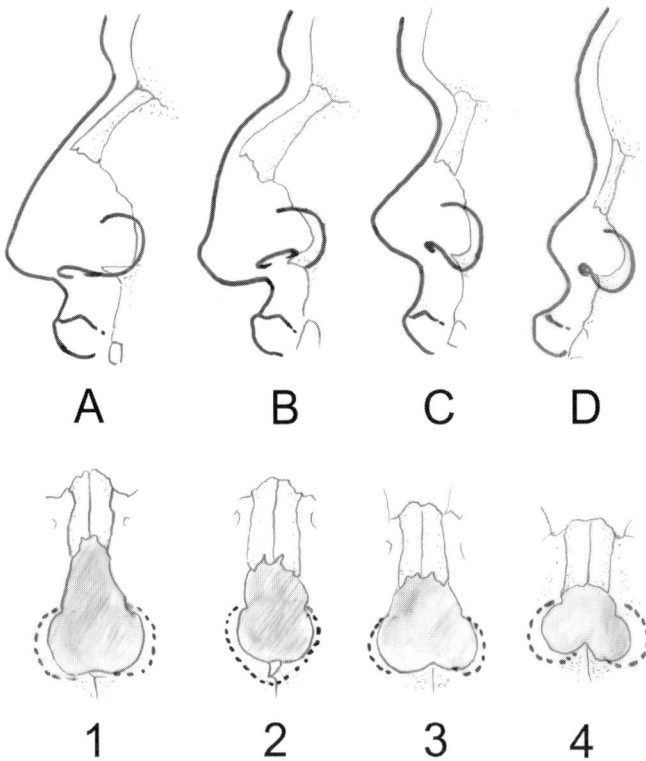

Fig. 4.9 The correlation between the soft and bony nose. A = straight nose, B = hawk-like nose, C = snub nose, D = upturned nose. 1–4 show different nostril shapes and nasal aperture patterns. Modified from Gerasimov (1975).

alar height will be clearly apparent in the shape of the lateral nasal bones.

One study of 353 Whites (199 female, 154 male) using lateral radiographs found that age was more important in determining the height and length of the nose in males than it was for females (Macho, 1986). The nose changed its appearance with age and the nasal septum tended to sink downwards with age. There was a negative correlation between the minimum thickness of the soft tissues along the nasal bones and the prominence of the bony nose, indicating that more prominent noses have a thinner layer of tissue than less prominent ones. Over the nasion it was revealed that high and long noses have a thinner soft tissue cover than the short noses. These results suggested that the profile line tends to adjust to disharmonies in the bony profile, and the length and height of the nose can be predicted with some certainty. McClintock Robinson *et al.* (1986) carried out a similar cephalometric radiograph study of 123 White females and found that people with straight profiles tend to have straight noses, convex profiles have convex nasal shapes and concave profiles have concave nasal shapes. George (1987) carried out a study of the noses of 54 American Whites (17 male and 37 female) using lateral cephalographs. He stated that since the nose is such a distinguishing feature of the face, it is indeed frustrating that the bony framework is so limited. Nonetheless, he stated that it was possible to reconstruct an average nose to fit the skeletal outline using his method. Using three case studies, his method of two-dimensional reconstruction was assessed and found to be accurate.

General rules for the reconstruction of the nose have been developed using the results of these and other studies (Goldhamer, 1926; Genecov *et al.*, 1990). In 1957, Broadbent and Mathews stated that the bony structure of the nose is half the nasal length, the nasal wings extend to the inner canthus and the profile of the nose is parallel to the axis of the ear. Angel (1978) stated that the form of the nasal bones must determine the nasal shape and that the slope of the nasal spine reflects the slope of the nasal base (an upward-sloping spine leads to an upturned columella) and the form of the nasal spine determines the nasal tip (a spatulate spine leads to a wide or bulbous tip, a bifid spine leads to a cleft nose etc.). He stated that the lateral spread of the piriform aperture sets the nostril width and that the nostrils bulge 2–3 mm beyond the lateral piriform edge. Clearly asymmetries must be taken into account. Gatliff and Snow (1979) stated that nasal width was equal to the nasal aperture width plus 10 mm in Whites, and plus 16 mm in Blacks, and the nasal projection was three times the length of the nasal spine. Taylor (2001) stated that there is a relationship between the canine fossa and the nasal width, in that the root of the canine tooth is directly below the widest nasal point. Gerasimov (1971) provided the best guide to nasal projection (see Fig. 4.10), and stated that '*the profile of the nose is projected by two straight lines,*

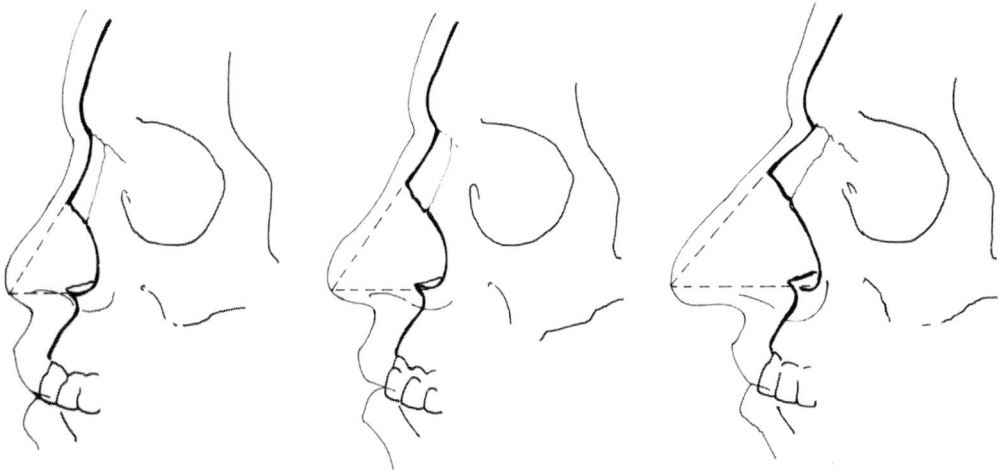

Fig. 4.10 The angle of the nasal spine determines the angle of the columella. Illustration by Richard Neave.

one at a tangent to the last third of the nasal bones, and the other as a continuation of the main direction of the point of the bony spine. The point of intersection of these two lines will generally give the position of the tip of the nose.' But he warned that it would be wrong to think that the first line would correspond to the profile of the nose, and stated that the line of the profile of the external nose was defined by the degree of *'undulatedness'* and general character of the aperture seen in profile. Stewart's study in 1983 of the Terry collection of skulls and death masks from the Smithsonian Institute substantiated Gerasimov's standards for nasal form, projection and asymmetry. Gerasimov also stated that, in general, the width of the nasal aperture at its widest point was three-fifths the overall width of the nose. Krogman stated, in unpublished work, that the nasal tip is relatively pointed in Whites and rounded in Blacks. He also stated that Whites show a long, oval and oblique nostril shape and Blacks show a rounded nostril shape. These statements were substantiated by research from the Terry Collection of death masks at the Smithsonian Institute (1636 cadavers).

Suk (1935) attempted to determine the position of the nasion and the subnasale through palpation of the noses of 16 cadavers. Each nose was then dissected and the exact position of the nasion and subnasale was established. Measurements of nasal length, breadth and nasal aperture length and breadth were also taken. Suk concluded that the bony nose does not tell us anything at all about the external nose and that the location of bony landmarks from the surface anatomy is very inaccurate and should be abandoned. Lebedinskaya *et al.* (1993) collected a considerable number of x-rays from Russians. They studied the nasal shape and found that the highest correlations were between the height of the bony nose and the general length of the nasal ridge, and the height of the nasal aperture and length of the cartilaginous nose. They also found that there was no relationship between the nasal width and the nasal aperture width, but that there was a strong correlation between the width of the nose and the distance between alveolar protrusions of the canines. Fedosyutkin and Nainys (1993) further stated that an exposed nasal septum was characteristic of a crest-shaped nasal base. Gerasimov (1971) produced a detailed guide to nasal shape. He stated that a straight, thin nose usually has a weak glabella, a narrow interorbital distance, a narrow and high nasal bridge, elongated narrow aperture with simple contours, a high roof-like dorsal part of the nasal bones, sharp lower angles of the aperture, a nasal spine which is never directed downwards and is usually directed forward, thin nasal bones and a sharp facial profile. A hooked or hawk nose was described as being characteristic of a face with a strong glabella, sharp and prominent nasal bones, a narrow base, thick bell-shaped nasal bones, a symmetrical aperture, a nasal spine which is strongly developed and directed forwards or downwards (when the spine is directed downwards the nose is bent and bill-like), and a narrow face. A fleshy broad nose was described as being characteristic of a broad roundish aperture, broad rounded nasal bones with simple contours and a bell shape, forward or upward-directed nasal spine, and a hearth-like lower border. He also stated that the snub nose usually has a short aperture, dull and rounded lower border, short upward-directed nasal spine and a concave roof with a wide and rounded arch. Gerasimov suggested that any nasal asymmetry will be present in the external nose. There is general agreement with this statement, although asymmetry in the ethmoid bone is an exception.

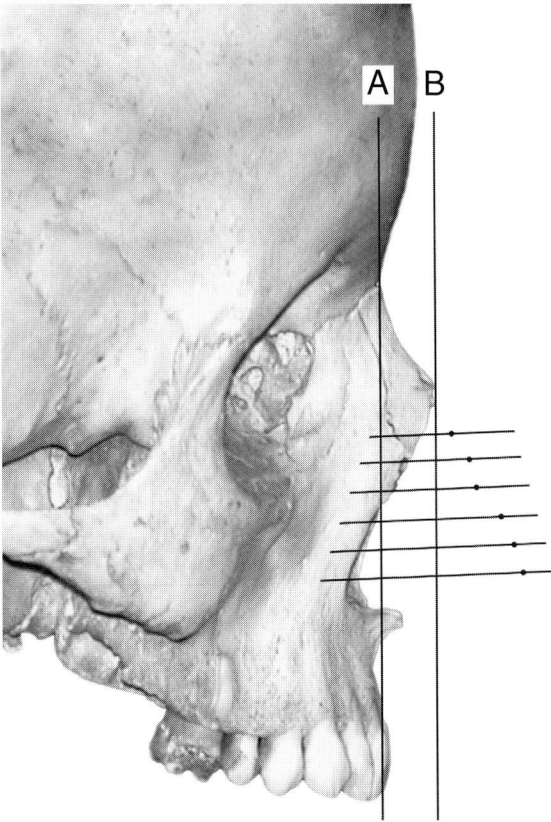

Fig. 4.11 The profile of the nose as determined by the Lebedinskaya method. Line A dissects the nasion and prothion. Line B is parallel to line A and intersects the foremost point on the nasal bone. Four to six equidistant lines were drawn perpendicular to these line between the nasal bone and the base of the piriform aperture. At each perpendicular line the distance from line B to the piriform rim was measured and the same distance added on the other side of line B. Using this method the profile of the nose was predicted. Modified from Prokopec and Ubelaker (2002).

The most recent research was carried out by Prokopec and Ubelaker (2002) at the Institute of Ethnology and Anthropology, Moscow. They studied four skulls and reconstructed the noses two-dimensionally using Lebedinskaya's method. An image of the skull in profile was produced (see Fig. 4.11) and a line (A) drawn dissecting the nasion and prothion. Another parallel line (B) was drawn intersecting the foremost point on the nasal bone, and four to six equidistant lines were drawn perpendicular to these lines, between the nasal bone and the base of the piriform aperture. At each perpendicular line the distance from line B to the piriform rim was measured and the same distance added on the other side of line B. Using this method the profile of the nose was predicted.

The eyes

The eyes are a vital part of the complete face, and their position clarifies the correct proportions of the middle third of the face. Stewart (1983) stated that Whitnall carried out research in 1911 at Oxford, where he published the first known description of the malar tubercle. Wilder's description (1912) was '*The position of the two canthi is almost precisely determined, the inner by the naso-lacrymal duct (lacrimal fossa) and the outer by the slightly but definitely indicated malar tubercle.*' Whitnall studied 2000 skulls from 23 racial groups and failed to find the malar tubercle on only 4.5 per cent of the skulls. Stewart (1983) carried out further research into the position of the palpebral ligaments at the inner canthus on the Terry Collection of skulls. He found that the anterior surface of the lacrimal sac best represented the level of the medial angle of the eye fissure. Van den Bosch *et al.* (1999) studied the eyes of 320 subjects and discovered that the eye fissure lengthens by 10 per cent between the ages of 12 and 25, and shortens by almost the same amount between middle age and old age. Aging causes sagging of the lower eyelid and a higher skin fold and eyebrow position, but does not alter the position of the eyeball or of the lateral canthus. Broadbent and Mathews (1957) stated that the upper eyelid is heavier and overlaps and extends lateral to the lower eyelid at the lateral canthus. Angel (1978) places the inner canthus 2 mm lateral to the lacrimal crest and at its middle, whilst the outer canthus is placed 3–4 mm medial to the malar tubercle (see Fig. 4.12). He also stated that a low nasal root plus a strong anterior lacrimal crest suggested the presence of the medial epicanthic (Mongoloid) fold, and a low orbit with an overhang of the brow ridge suggested an intermediate plus a lateral epicanthic fold. Also, a strong posterior lacrimal crest suggested a wider intercanthal breadth. However, the most detailed study of

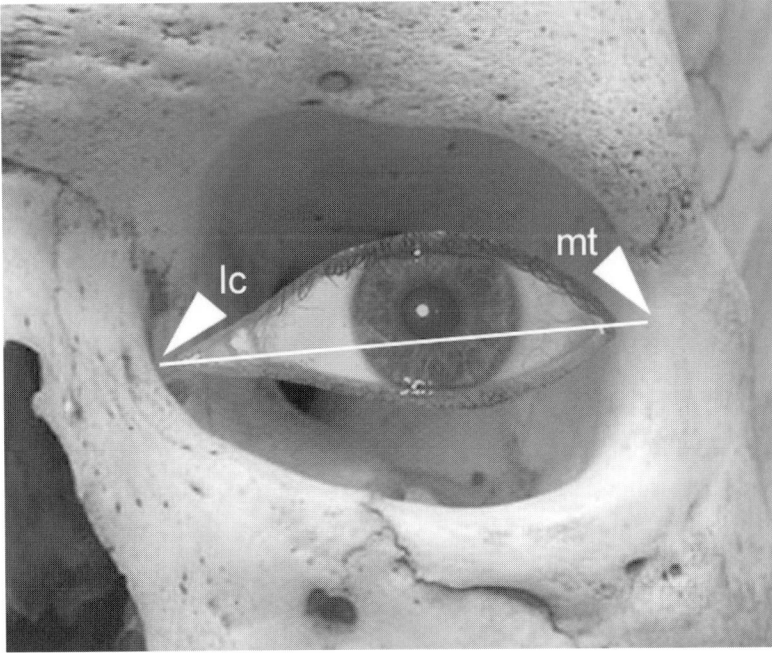

Fig. 4.12 The position of the lacrimal crest (lc) and malar tubercle (mt) at the orbit.

the eyes was by Fedosyutkin and Nainys (1993), who found that the length of the eye fissure was 60–80 per cent of the width of the orbit. When the malar tubercle was absent (in 15 per cent of skulls) they suggested placing it 8.5 mm below the frontozygomatic suture in males and 9.5 mm below in females.

Fedosyutkin and Nainys also found that the folds of the eyelid could be determined by the structure of the supraorbital rim (see Fig. 4.13). The fold of the fixed part of the upper eyelid generally followed the direction of the supraorbital rim. However, if there was an overhang in the middle of the supraorbital margin, the fold of the eyelid was located in the middle of it. If the lateral rim was thickened and slanted upwards and posteriorly, the fold was more laterally pronounced. When there was a high orbit, a low or medium-height nasal root and a long lacrimal fossa, the eyelid fold was more medially pronounced. Wolff (1997) stated that the anteroposterior eyeball diameter was 24 mm in adults, with a mean

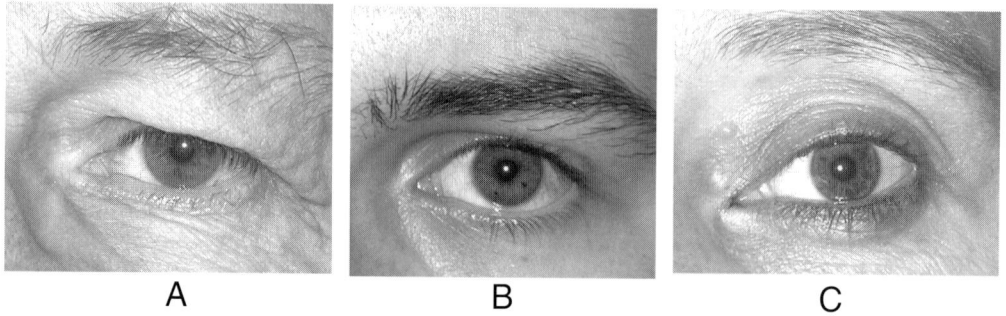

Fig. 4.13 Eyelid patterns. Epicanthic fold positioned medially (A), centrally (B) and laterally (C).

male diameter of 24.6 mm and a mean female diameter of 23.9 mm. He found no differences in eyeball size related to racial origin. Other studies are in agreement with Wolff (Stenstrom, 1946; Gray, 1980; Tian *et al.*, 2000). However, Gerasimov (1971) discovered that Caucasoids exhibited the largest eyeballs, Negroids had medium-sized eyeballs and Mongoloids had the smallest eyeballs. He found that Negroid skulls had the smallest orbits, Caucasoid skulls medium-sized orbits and Mongoloid skulls the largest orbits. Therefore, he concluded that the size of the eyes was not correlated to the diameter of the orbits. Eyeball size has been shown to be relatively constant throughout adult life (Imai & Tajima, 1993; Furuta, 2001).

The protrusion of the eyeball in the socket has been an area of some research. Wolff (1997) stated that '*a straight line placed against the superior and inferior orbital margins will just touch or miss the front of the cornea*', and Gatliff and Snow (1979) position the eyeball in the orbit by following this standard. Fedosyutkin and Nainys (1993) stated that the protrusion of the eyeball should be determined by the depth of the orbital cavity, vertical inclination of the orbit, and the thickness and degree of overhang of its upper rim. Deep-set eyes were found more often when the upper orbital rim was greatly thickened and protrusive, relative to the lower one. A wide opening of the eye fissure and protruding eyeballs were characteristic of a weak orbital profile and smooth thickened outer rim as it reached the malar tubercle. Research has shown that normal eyeball position varies in an anterior–posterior direction by as much as 3.7 mm (Bogren *et al.*, 1986). Wilkinson and Mautner (2003) studied the MRI scans of 39 subjects and found that the mean eyeball protrusion was 3.8 mm more anterior than Wolff's standard. They suggested that the eyeball should be positioned so that the tangent between the superior and inferior orbital margins touches the iris, rather than the cornea. This standard takes into account Wolff's research (1997) that suggested the cornea was approximately 3.8 mm anterior to the iris. Further research (Stephan, 2002) studied previous exophthalmometry papers (Bogren *et al.*, 1986; Yeatts, 1992) to collate data on eyeball projection measurements. Stephan was in agreement with Wilkinson and Mautner, and found that the cornea was approximately 4 mm anterior to the suprainfraorbital tangent. Wilkinson and Mautner (2003) also found that orbit depth increased as eyeball protrusion decreased.

Fedosyutkin and Nainys (1993) also stated that the eyebrow pattern could be determined from the supraorbital bony structure. If there was strong development of the supraorbital margin and brow ridge, the eyebrows were shifted downwards, 1–2 mm lower than the supraorbital margin. A weakly-developed nasal bridge and brow ridge suggested that the medial third of the eyebrow was located in the orbit, just beneath the supraorbital margin, whilst the lateral two-thirds of the eyebrow gradually rose to the supraorbital margin and traced its contour. Finally, if there was thickening of the lateral part of the supraorbital rim and a strong brow ridge, the eyebrow was arranged over it in a triangular form (see Fig. 4.14). Angel (1978) stated that the eyebrows generally followed the line of the brow ridge and that they were approximately 3–5 mm above the supraorbital margin. Angel's standards were confirmed by Stewart's study (1983) of the Terry Collection of skulls and death masks. Gerasimov (1971) stated that the muscles surrounding the eyes were closely co-ordinated in their shape and form and degree of development with the relief of the orbits, their general shape and position, the degree of projection of the eyeballs and their position. He stated that '*it is only by taking into account all these details that a correct reproduction of the outer form of the eye is possible*'.

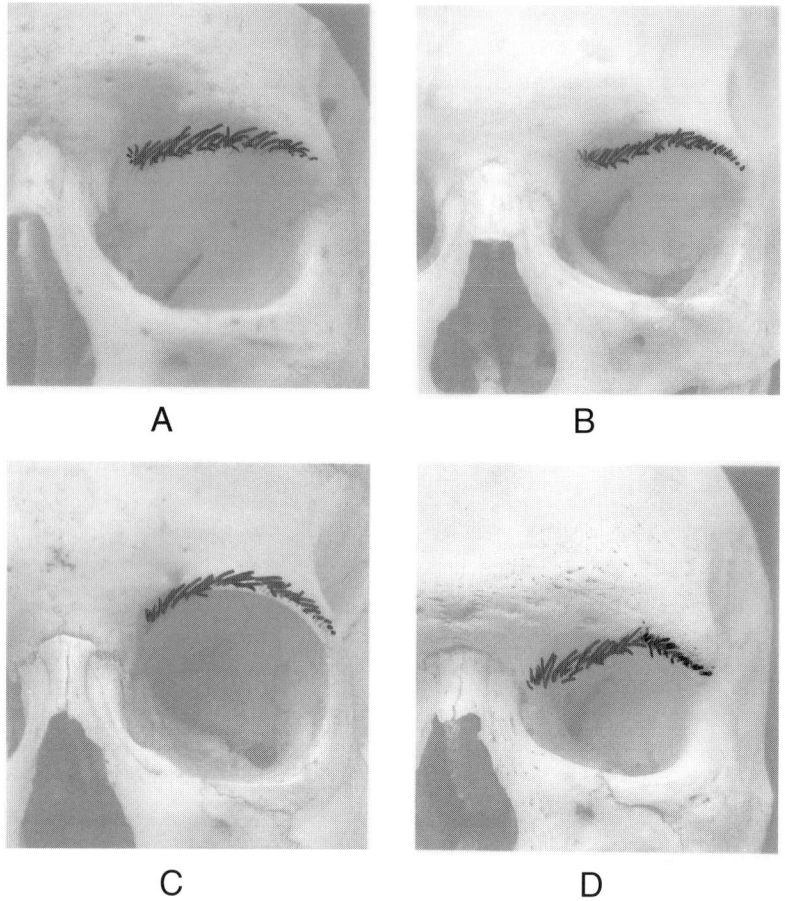

Fig. 4.14 Eyebrow patterns. A = strong supraorbital margin and brow ridge. B = weak brow ridge and low nasal root. C = weak brow ridge and high nasal root. D = thickened lateral supraorbital rim and strong brow ridge. Modified from Fedosyutkin and Nainys (1993).

The mouth

The lips and the mouth are very important areas of the face with regard to appearance, and research suggests that some detail of these facial features can be determined from the skeletal structure. Broadbent and Mathews (1957) stated that the junction of the upper and lower lips is on a line perpendicular to the medial border of the iris. Gerasimov (1971) stated that the thickness of the

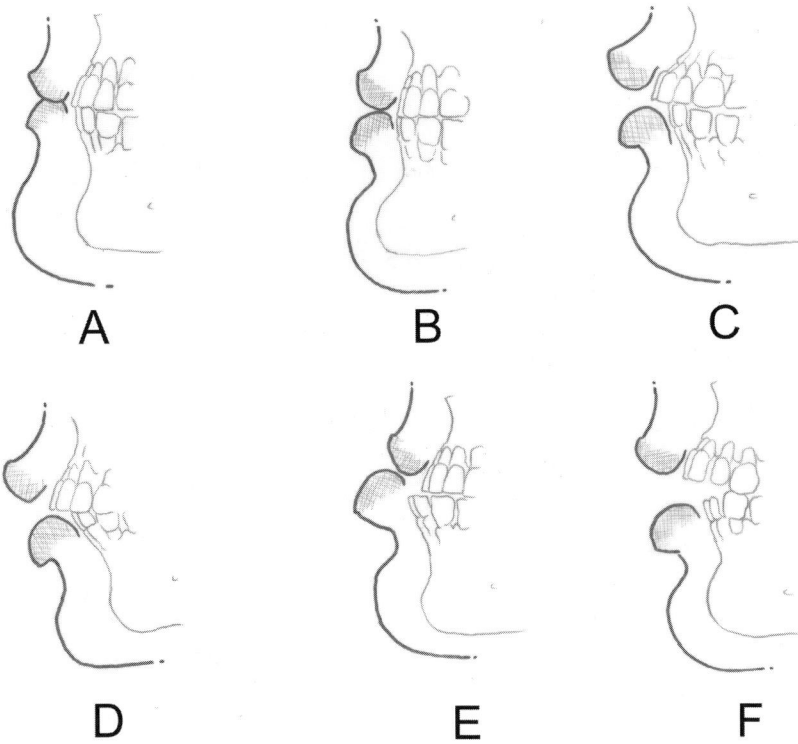

Fig. 4.15 The basic mouth forms. A = step, B = pincer-like, C = scissor-like, D = roof-like, E = cornice-like, F = open occlusion. Modified from Gerasimov (1975).

lips is based on the prognacy of the teeth, the incisors, and alveolar parts of the upper and lower jaws (see Fig. 4.15). Small straight teeth are characteristic of thin lips and orthognathism. Prominent big teeth are characteristic of thick lips and prognathism. He also stated that the height of the enamel of the middle incisor is equal to the thickness of the middle of the pigmented part of the lip, but Gerasimov warned that the thickness of the lips is not a constant feature, that it changes with age and varies within the same racial group. He also stated that the mouth form can be determined from the occlusion of the teeth, the dental pattern, the morphology of the jaw and the facial profile. He stated that prognathism in Negroids, Polynesians, Japanese and Malays would suggest thick lips, whereas prognathism in Caucasoids or Mongoloids is not accompanied by procheilia of the lips, and the mouth has a puckered appearance with the upper lip leaving the teeth uncovered at rest. He further stated that mandibular prognathism suggests procheilia of the lower lip, and maxillary prognathism suggests procheilia of the upper lip.

Angel (1978) found that the line of the lower edge of the lower lip is just above the middle of the incisor crowns, and the corners of the mouth are found at the first premolar–canine junction. Stewart (1983) substantiated Angel's standards in his study of the Terry collection of skulls and death masks. Angel also stated that the lip thickness depends upon the projection of the teeth, racial group, and the strength of incisive and buccinator muscles. Gatliff and Snow (1979) placed the corners of the mouth by aligning them with the midpupil and the widest point of the chin. However, Krogman and Iscan (1986) placed the corners of the mouth directly below the centre of the pupils. Latta et al. (1991) studied the relationship between the width of the mouth, interalar width, bizygomatic width and interpupillary distance on 109 edentulous patients, and Wilkinson et al. (2003) studied mouth width, interpupillary and interlimbus distances on 96 young adults. These studies found no correlation between mouth width, bizygomatic width, interpupillary and interalar distances. Wilkinson et al. (2003) found that the interlimbus distance was the most accurate determinant of mouth width, and this relationship was not affected by sex or ethnic group. It appears, therefore, that the most accurate placement for the corners of the mouth are on a radiating line from the first premolar–canine junction and/or the medial border of the iris. Wilder (1912) stated that the mouth slit, when the mouth is in repose, coincides with the occlusive line of the teeth. This seems in disagreement with the majority of other sources (Gerasimov, 1975; George, 1993; Greyling & Meiring, 1993; Ferrario et al. 2000) who state that the oral fissure cuts across the lower third or quarter mark of the maxillary central incisor (see Fig. 4.16).

Gatliff (1984) determined the vertical thickness of the lips from the gumline-to-gumline measurement. Recent studies by Wilkinson et al. (2003) suggested that there is a positive correlation between

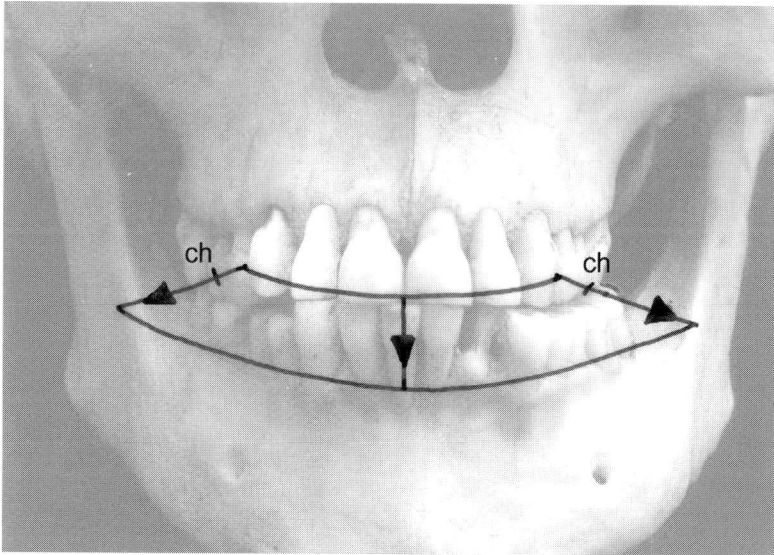

Fig. 4.16 The position of the oral fissure as determined by the teeth. ch = chelion points (corners of the mouth).

the upper lip thickness and maxillary teeth height, lower lip thickness and mandibular teeth height, and the total lip thickness and total teeth height, but suggested different standards dependent upon ethnic group. They suggested that for White Europeans, lip thickness can be calculated from teeth height by the following formulae:

upper lip thickness = $0.4 + 0.6 \times$ (upper teeth height)
lower lip thickness = $5.5 + 0.4 \times$ (lower teeth height)

Lip thickness can be calculated from teeth height, for Asians from the Indian subcontinent, by the following formulae:

upper lip thickness = $3.4 + 0.4 \times$ (upper teeth height)
lower lip thickness = $6 + 0.5 \times$ (lower teeth height)

Krogman and Iscan (1986) stated that both Whites and Blacks will have a well-developed philtrum and will show a cupid's bow-shape to the upper lip. He claimed that in Whites the vermilion shows a smooth junction with the upper lip whereas in Blacks it is elevated to give what Krogman described as the 'lip seam'.

Fedosyutkin and Nainys (1993) studied the teeth–lip relationship and suggested that the philtral width corresponded to the distance between the midpoints of the upper central incisors. They stated that if the teeth show an overbite, or maxillary prognathism, then the upper lip will project further than the lower lip, and the lower lip will project further than the upper lip if there is edge-to-edge occlusion or an underbite. Fedosyutkin and Nainys concluded by saying that the remaining characteristics of the mouth and lips are derived by an 'educated guess'. Latta (1988) studied the midline and its relation to anatomical landmarks in 100 subjects. He found that the width of the mouth (mean, 52 mm) and the width of the philtrum (mean, 11 mm) were correlated irrespective of age, sex or racial group. He also found that neither of these distances was affected by age and that the size of the mouth and philtrum were larger for men than women, and larger for Blacks than Whites.

Angel (1978) suggested that the relative strength of markings for the levator and depressor anguli oris muscles determines the up or down placement of the corners of the mouth. The markings for the origins of the levator labii superioris and zygomaticus muscles determine the curves and depth of the nasolabial fold and the possibility of a second lateral crease. The nasolabial fold was described by Fedosyutkin and Nainys (1993) as extending from the upper edge of the nostril toward the upper first molar. They found that its strength depended on the depth of the canine fossa, the degree of horizontal face profiling, the projection of the frontal surface of the cheekbones and the presence or absence of the teeth. They stated that the nasolabial fold is more pronounced when the canine fossa is deep (more than 5 mm), also when the teeth are missing and with advancing age. Gerasimov (1971) suggested that the nasolabial fold starts from the lateral border of the aperture above the crista conchalis, and passes through the middle to terminate just below the second molar in the direction of the inner angle of the lower jaw. He stated that the interrelation of these points defines the contour of the nasolabial fold.

The chin

Tandler (1909) found that the degree of chin protrusion and the thickness of soft tissues at this part of the face were not connected. Fedosyutkin and Nainys (1993) found that the degree of elevation of the frontal part of the mandible and the width of its base define the width of the chin, and that if the height of the mandibular body decreases from the chin triangle to the side of the rami, then this forms a high chin. Gerasimov (1971) found that when the lower border of the jaw is softly rounded inwards, has no crests and possesses no roughness, then the muscular tissue will softly cover the bone, and the chin will have gentle contours. He suggests that when the lower border shows prominent crests then, correspondingly, there will be well-developed muscle and the chin will be heavy and massive in form. A cleft chin will be suggested by a central groove on the mental eminence and strong muscle attachments for the mentalis muscles.

The ears

There is some dispute over the axis of the ear placement. Traditionally, anthropologists suggest that the axis of the ear (from the most superior point on the helix to the most inferior point on the lobe) is parallel to the nose (from tip to nasion) (Broadbent & Mathews, 1957; Gerasimov, 1971). However, this theory has no basis in scientific study and the most convincing theory is from Skiles and Randall (1983). They studied 46 subjects and found that the angle between the axis of the ear and the angle of the nose was 15°, and that this was significantly different from the parallel. On top of this, observers were asked to choose the most aesthetically pleasing ear orientation from a group of four possibilities, and overwhelmingly chose an orientation with the helix more anterior than the parallel. Preeyanont (1995) studied 619 Thai women and found that the longitudinal axis of the ear was not parallel to the bridge of the nose, and that the angle of the ear should be 19.3° to achieve the most aesthetically pleasing position. However, Preeyanont stated that the ideal angle probably differs, depending on ethnic origin. Farkas *et al.* (2000) also found that a greater inclination of the nasal bridge than that of the longitudinal axis of the ear was a very frequent canon variation in both African–American and North American Caucasian populations. Another study (Borman *et al.*, 1999) on Turkish subjects agreed and showed that only 3% of the population had equal ear and nose inclination, whilst 88% had a greater nose than ear inclination. Gatliff stated that the external auditory meatus should be placed at the top of the tragus of the ear. Some experts (Gatliff & Snow (1979); Charney & Coffin (1981); Fedosyutkin & Nainys; (1993)) suggest that the height of the ear is approximate to the length of the nose. Angel (1978) stated that the size, form and projection of the ears are hard to estimate, and Gatliff (1984) agreed that there are very few clues about the exact shape of the ears from the skull. However, Fedosyutkin and Nainys (1993) suggested that considerably more information about the ears can be determined from the skull. They found that the mastoid processes may determine some of the features of the ears and state that if the supramastoid crest on the temporal bone is strongly developed and protrudes, then the ear will show upper protrusion. If the outer surface of the mastoid process is rough then the ear will show lower protrusion. If all these attributes are in place, then the ear has total protrusion. They further suggested that if the mastoid processes are directed downward (with the skull in the Frankfurt Plane), the lobe is attached (adherent), and if the mastoid processes point forward, then the lobe is free (non-adherent). Gerasimov (1971) agreed that much information regarding the ear can be suggested by the skull details, and stated that the breadth of the ear will be approximate to half its length and the axis of the ear is parallel to the axis of the jaw line (see Fig. 4.17). He also stated that small, medially-directed mastoid processes will suggest small ears that are close to the head; whilst massive, prominent mastoids are characteristic of large spread-out ears. None of these theories has documented research as a basis for

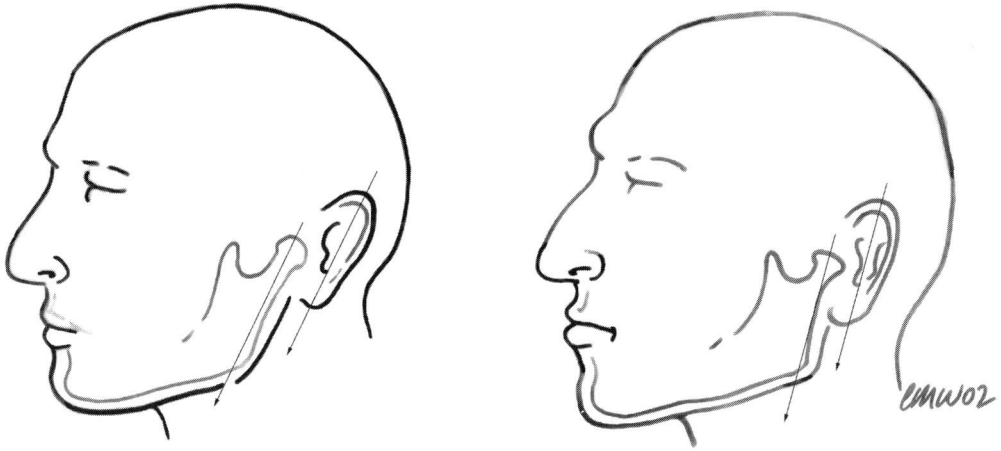

Fig. 4.17 The angle of the ear as determined by the jawline.

the conclusions. Caution should be exercised when reconstructing the ears and any attempt at more than a broad description should be avoided. The importance of ear size and shape in facial recognition is unclear. As Gerasimov summarised, '*many details of the ear's complicated relief must be intuitively reconstructed*'.

One of the factors that can have a great effect on facial appearance is stature, and where possible this factor must be taken into account. Evidence of emaciation or obesity may be seen on the skeletal remains. For example, if a person has been obese for a significant amount of time, then the muscles of the face would have been carrying more weight and one would expect to find deeper, more defined muscle attachments on the surface of the bone, especially at muscles such as masseter, temporalis, sternocleidomastoid, the zygomatics and the levator muscles of the mouth. Similarly, with emaciation that has been present for a significant amount of time you would expect to see signs of malnutrition and muscle wastage. However, within the normal range of stature it is impossible to predict facial fatness from the skeletal detail, and the remains of clothing can be the only useful clue. Gerasimov (1971) found that '*different races usually possess different thicknesses of the soft parts of the face. But these differences are also constitutional and it may be that the constitution of the subject plays a more important role than race.'* A forensic facial reconstruction must rely upon the assumption that those people who know the individual well, such as family members or close friends, would recognise the individual even with the loss or gain of a couple of stone.

There have been many critics of facial reconstruction who claim that, although clearly related to one another, it is impossible to predict the soft tissues of the face from the skeletal structure. Suk (1935) was critical of facial reproducibility from the skull because of the in vivo appearance of a face with respect to muscle tone, emotion etc. He used the example of portraiture to make the point that 'candid camera' pictures are often better examples of the individual, with regard to recognition, than a portrait photograph, which may include an unrelaxed, formal pose or indeed a too relaxed, dream-like pose. He also indicated the undoubted difference between the animated face of a person in life with the same face in death. He therefore suggested that it is a great mistake to think that the features of the face depend on the bony structure of the skull and concluded that a facial reconstruction from the skull must resort to fantasy. Brues (1958) stated that facial reconstruction is *'probably best left to the ample literature of detective fiction'*. However, the research discussed in this chapter suggests that all features of the face are determined by the skeletal shape and form. Obviously the location, size and shape of the features are important to the individualisation of the face and must be carefully measured and located. Krogman and Iscan stated in 1986 that *'the skull is the matrix of the living head; it is the bony core of the fleshy head and face in life'*. Gerasimov (1971) claimed that the main mistake made when carrying out a facial reconstruction was to view the separate details of the face as something independent or isolated from the general composition of the face. He suggested that the facial form should be recreated using observations on the correlation between muscle attachments on the skull and the degree of size and configuration of the separate parts of the skull. The caricaturist Drucker (George, 1993) stated, *'We all have the same features, it's the spaces between them, their proportions and relationships to one another that distinguish one face from another.'* However, I am in agreement with some authors such as George (1993) who suggest that there are many facial variations and, since faces age at different rates and with different intensities, and the nutritional status of the face is unpredictable, any facial reconstruction cannot create a portrait – more likely it can create an approximation based on the proper alignment of features and the characteristic details of the face. Some level of individuality can be derived from the skull (Fedosyutkin & Nianys, 1993), but the accuracy of reproducing attributes of appearance depends on the amount of detail obtained from the forensic examination of the remains by an experienced expert.

5 Facial tissue depth measurement

The variation in facial tissues between sexes, ages, ethnic groups and different nutritional states has been studied for the last 120 years, and this chapter attempts to collate all the available data. Knowledge of common facial anthropometry terms is useful when assessing these studies, as a wide variety of anatomical points have been employed. The number and position of the points included for each study of facial tissue depth are determined by the researcher, and commonly they range between 15 and 34 points. Some researchers use different terms for the same points, and Fig. 5.1 illustrates these terms.

v	vertex	The highest point on the head.
of	ophryon	The most anterior midline point of the forehead.
g	glabella	The most prominent midline point between the eyebrows, identical to the bony landmark.
op	opisthocranion	The point on the occipital region that is furthest from the glabella.
fe	frontal eminence	A point between the metopion and the ophryon on the lateral frontal bone, directly above the midpoint of the eyebrow.
ft	frontotemporale	Lateral point from the elevation of the linea temporalis, or the terminal points of the tail of the eyebrow.
tr	trichion	Midpoint of the hairline.
zy	zygomatic arch	Most lateral point over the zygomatic arch, identical to the bony landmark. Also known as the zygion.
go	gonion	The most lateral point on the mandibular angle, close to the bony gonion.
lm	labiomental	The midpoint of the labiomental groove. Also known as the sublabiale or supramentale point.
m	mental	The most anterior midpoint of the chin, on the surface above the bony landmark. Also known as pogonion.
mn	menton	The lowest medial landmark beneath the chin. Also known as the bony gnathion.
en	endocanthion	Inner corner of the eye.
ex	exocanthion	Outer corner of the eye.
or	infraorbital	The lowest point on the lower margin of the orbit. Also known as orbitale and suborbital.
os	supraorbital	The highest point on the upper margin of the orbit.
lo	lateral orbit	A point on the lateral orbital margin on a line with the eye fissure. Also known as the ectoconchion.
sci	superciliary	The highest midpoint of the eyebrow.
n	nasion	The midline point of the nasal root, identical to the bony landmark.

ns	midnasal	The midpoint of the nasal bones.
rh	rhinion	The end of the nasal bones at the cartilage–bone junction.
ln	lateral nasal	A point on the lateral nasal bone that is on the Frankfurt Plane.
al	alare	The most lateral point on each alar contour. Also known as *supracanine*.
prn	pronasale	The most protruding point at the tip of the nose.
sn	subnasale	The midpoint of the columella base below the nasal spine.
cf	canine fossa	The deepest point at the canine fossa. Also known as *maxillary point*.
im	inferior malar	Beneath the zygomatic attachment point on a vertical line with the infraorbital and supraorbital points.
mp	midphiltrum	The midpoint of the philtral column.
ls	upper lip border	The midpoint of the upper vermilion line.
li	lower lip border	The midpoint of the lower vermilion line.
sto	stomion	The midpoint of the incisive line of the lips.
ch	chelion	The corner of the mouth fissure.
sa	superaurale	The most superior point on the helix of the ear.
sba	subaurale	The most inferior point on the helix of the ear.
po	porion	The highest point on the upper margin of the external auditory meatus.
sg	supraglenoid	The point anterior to the tragus at the root of the zygoma. Also known as the *pretragal* point or the *root of the zygoma*.
mm	midmasseter	A point on the occlusal line at the centre of the ramus of the mandible. Also known as the *occlusal line*.
mdm	midmandible	A point on the mandibular border inferior to the premolar region.
za	zygomatic	Point of the zygomatic muscle attachments on the cheek bone. Also known as *malare*.

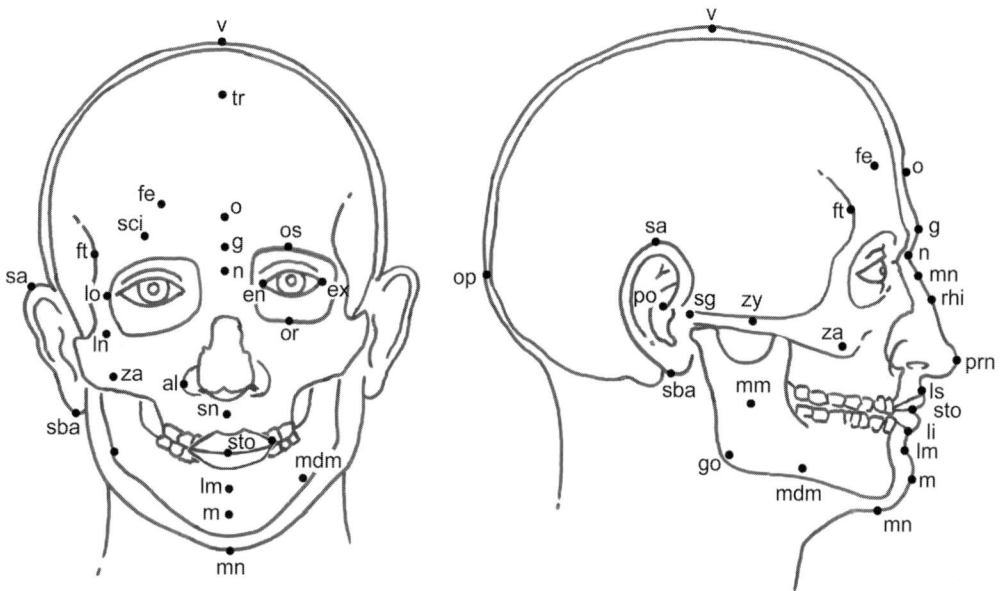

Fig. 5.1 Common facial anthropology points. Modified from Helmer (1984).

History

The first research into facial tissue depths was as an accompaniment to the facial reconstruction technique. Welcker carried out the first documented research in 1883. He studied 13 white male cadavers of middle age, and measured facial tissue depths using a double-edged, chisel-shaped knife blade pushed into the flesh of the cadaver. Welcker established tissue thicknesses at nine midline points. His, the German anatomist, studied 24 male and four female White cadavers in 1895 in Leipzig, using a sewing needle with a rubber disc attached. The needle was pushed into the flesh and the motion pushed the rubber disc along the needle to show the depth entered (see Table 5.1). His studied nine midline facial points and six lateral points. Kollman and Buchly (1898) expanded on His's research and added three new sites (see Table 5.1). They studied 21 male and four female White European cadavers, between the ages of 17 and 72 years, and developed a new method, which employed a needle covered in soot. Following cadaver flesh penetration, the clean area of the needle indicated the thickness of the tissue. Czekanowski (1907), Berger (1965) and Leopold (1968) also measured facial tissues from Caucasian cadavers (Helmer, 1984). These Caucasian cadaver measurements are summarised at Table 5.2. Sutton (1969) measured the thickness of soft tissue over the zygomatic attachments in 104 Caucasian cadavers over three body-builds (thin, medium and fat). Sutton used the His technique of measurement. He found that the measurements were not significantly different between the two sides of the face, and that the variation was very large (1.4 mm to 21.4 mm) with a marked relationship between body-build and tissue thickness of the face (male: 8 mm thin, 12 mm medium, and 21 mm fat; female: 10 mm thin, 15 mm medium, and 21 mm fat). Suk (1935) criticised the needle-puncture technique, due to the impossibility of ascertaining accurately the different landmarks for measurement because of human error.

During the twentieth century, anatomists studied facial tissue depths in many varied racial and ethnic groups. Birkner (1903–1907) studied Chinese cadavers, Fischer (1903) studied Papuan

Table 5.1 Early facial tissue measurements from White European cadavers.

Facial points	His (1895) Well-nourished					Emaciated		Kollman & Buchly (1899) Well-nourished			
	Emaciated Male (9)	Male (24)		Female (4)				Male (21)		Female (4)	
	Mean	Mean	Range	Mean	Range	Male Mean	Female Mean	Mean	Range	Mean	Range
Forehead	3.4	4.1	3–5	4.2	4–4.5		1.9	3.1	2–4	3.0	2–4.2
Glabella	3.9	5.2	4–6	4.8	4–5.5	3.0	2.9	4.3	3–5.8	3.9	3.2–5.4
Nasion	4.8	5.5	4–7	5.0	4.5–5.5	3.1	3.5	4.3	3–6	4.1	2.5–4.7
Midnasal bone	3	3.3	3–3.5	3.0	2.5–3.5	2.5	2.1	3.1	2.1–5	2.6	2–4
Rhinion						2.1	1.5	2.1	1.3–3	2.1	1.6–3
Subnasale	10.8	11.3	9–14	9.8	8–11	14.7	7.1	11.7	8.3–14.7	10.1	8–11
Midphiltrum	8.2	9.4	8–12	8.3	6–10	11.0	6.2	9.5	6.1–13	8.1	7–10
Labiomental	8.5	10.0	8–14	9.8	7.5–11	8.8	7.2	9.8	8–13.5	11.0	7.8–14.1
Mental tubercle	8.5	11.1	8–15	10.8	9–13	5.7	5.0	9.0	5–13	9.4	7.7–12.1
Menton	4.1	6.2	4–8	6.5	6–7	5.1	3.7	6.0	3–9	5.9	3.8–9.4
Supraorbital	4.6	5.8	4–8	5.5	5–7	3.8	4.1	5.4	2–6.8	5.2	4.6–5.5
Infraorbital	3.8	4.9	3.5–6	5.3	4–6	2.1	3.8	3.5	2.1–6.1	3.7	3–4.4
Zygomatic attach						3.2	4.2	6.6	3.2–10.9	7.7	6.7–9.5
Midmandible	4.8	8.4	6–12	8.1	7–8.5	5.0	3.6	7.8	2.3–12	6.2	4.7–8.5
Zygomatic arch	3.8	6.1	4–9	6.8	6–8	3.0	2.8	4.3	1.8–7.8	5.3	3.1–8
Midmasseter	13	17.6	11–22	17.0	14–19		11.5	17.0	6.3–24.5	14.8	12–19
Gonion	8	12.1	9–16	11.5	11–12	4.5	3.8	8.7	3–15.1	7.6	4.7–10.2
Supraglenoid						5.8	6.6	7.4	3.9–11	7.2	4.8–9.8

Table 5.2 Facial tissue measurements from White European cadavers.

Facial points	Welcker (1883) Male (13)		His (1895) Male (24)		His (1895) Female (4)		Kollman & Buchly (1899) Male (21)		Kollman & Buchly (1899) Female (4)		Czekanowski (1907) Male (64)		Czekanowski (1907) Female (51)		Berger (1965) Male (26)		Leopold (1968) Male (102)		Leopold (1968) Female (52)	
	Mean	Range	Mean	Range	Mean	Range	Mean	Range	Mean	Range	Mean	Range	Mean	Range	Mean	Range	Mean	Range	Mean	Range
Vertex	5.3	3–7									3.5	1.3–5.7	3.5	1.8–5.3			4.2	3–6	4	2.5–5
Forehead	4.3	2–5	4.08	3–5	4.16	4–4.5	3.07	2–4	3.02	2–4.2					4.1	2.8–6	5.8	4–7	4.9	3–7
Glabella			5.17	4–6	4.75	4–5.5	4.29	3–5.8	3.9	3.2–5.4	3.2	1.9–4.8	3.2	1.6–4.9	5.4	3.9–7.1	7	5–10	6.1	5–7
Nasion	5.9	3–9	5.45	4–7	5	4.5–5.5	4.31	3–6	4.1	2.5–4.7					5.7	3.5–7.6	3.8	2–6	3.4	2–5
Midnasal bone	3.3	2–5	3.29	3–3.5	3	2.5–3.5	3.13	2.1–5	2.57	2–4					2.7	1.7–4	2.4	1–3.5	2.3	1–4
Rhinion	2.2	1.4–3					2.12	1.3–3	2.07	1.6–3										
Subnasale	11	8–14	11.25	9–14	9.75	8–11	11.65	8.3–14.7	10.1	8–11							10.5	5–13	8.9	4–12
Midphiltrum			9.37	8–12	8.26	6–10	9.46	6.1–13	8.1	7–10							12.5	7–22	11	7–14
Upper lip																	10.9	8–14	10.8	8–14
Lower lip																	12.7	8–17	11.9	9–15
Labiomental	10.6	9–13	10	8–14	9.75	7.5–11	9.84	8–13.5	10.95	7.8–14.1					9.9	6.9–12.5	8.5	5–12	6.8	4–11
Mental tubercle			11.05	8–15	10.75	9–13	9.02	5–13	9.37	7.7–12.1					9.9	7.7–12.9				
Menton			6.16	4–8	6.5	6–7	5.98	3–9	5.86	3.8–9.4	3.1	1.2–6.4	2.7	1.4–4.3	6.8	4.3–9.6				
Supraorbital			5.8	4–8	5.5	5–7	5.41	2–6.8	5.15	4.6–5.5					5.9	4.1–8.4				
Infraorbital			4.9	3.5–6	5.25	4–6	3.51	2.1–6.1	3.65	3–4.4					4.1	1.8–6.1				
Zygomatic							6.6	3.2–10.9	7.73	6.7–9.5					8.6	5.9–11.2				
Lower 1st molar																	5.6	3.5–10.8		
Midmandible			8.37	6–12	8.1	7–8.5	7.76	2.3–12	6.16	4.7–8.5					8.3	5–11.2				
Zygomatic arch			6.05	4–9	6.75	6–8	6.62	1.8–7.8	5.32	3.1–8	3.2	1.1–6.9	3.8	1.1–6.5			9.2	6–12	7.7	5–11
Midmasseter			17.55	11–22	17	14–19	17.01	6.3–24.5	14.83	12–19					16.8	10.7–22.5				
Gonion			12.08	9–16	11.5	11–12	8.72	3–15.1	7.56	4.7–10.2	2.7	1.2–5.7	2.7	0.7–5.4	9.3	5.2–14.3	6.9	4–13	6.9	5–9
Supraglenoid							7.42	3.9–11	7.1	4.8–9.8					9.4	5.2–14.1				
Opisthocranion											4	2.6–7.6	3.1	1.6–7.1						

cadavers, Von Eggeling (1909) studied Herero (Namibian) cadavers, Suzuki (1948) studied Japanese cadavers, Rhine and Campbell (1980) studied American Black cadavers, Rhine *et al.* (1982) studied American White cadavers, and Rhine (1983) studied Southwestern Indian cadavers. The inter-racial cadaver tissue depth variation is summarised at Table 5.3.

There are some problems with cadaveric studies, the most obvious being the change in tissue depth due to loss of muscle tone and shrinkage. Soft tissue distortion occurs from drying and embalming even in the first few hours after death. Putrefaction, with bloating of the face, may occur rapidly even in temperate climates. The movement, skin elasticity and muscle tone in life all add bulk to a face, and the horizontal position of the cadaver when the measurements are taken create false tissue depth measurements due to the action of gravity. The results of these studies are not considered to be an accurate representation of the amount of tissue on the face in life. More recently Simpson and Henneberg (2002) studied the facial tissue depths of White Australian cadavers using the needle-puncture method (see Table 5.4). They compared the tissue measurements for unembalmed, recently embalmed (three months after death) and fully embalmed (6–12 months after death) cadavers, and found that embalming increased the soft tissue at all the facial landmarks. They also found that embalming initially increased the tissue thickness by approximately 55 per cent, but as the tissues became fully embalmed, the tissues decreased in thickness by approximately 20 per cent, due to dehydration. Simpson and Henneberg promoted the use of embalmed cadavers over unembalmed cadavers when measuring facial tissue.

Different and more accurate methods of measuring tissue depth were therefore studied. These methods included different measurements from imaging techniques such as craniographs, MRI scans and CT scans. The subject's head is placed in a standardised plane using a head positioner, and a radiograph is taken in the desired plane. Measurements can then be taken by hand from the developed x-ray. These imaging techniques suffer from a positional problem in that the subject must be in a supine position and the forces of gravity have an undeniable action upon the tissues of

Table 5.3 Comparison of facial tissues between cadavers from different ethnic groups.

Facial points	Combined[a] (1883–1968) White Europeans Male (58) Mean	Combined[a] White Europeans Female (8) Mean	Birkner (1906) Chinese Male (9) Mean	Fischer (1905) Papuans Male (2) Mean	Von Eggeling (1909) Hereros Male (3) Mean	Suzuki (1948) Japanese Male (7) Mean	Suzuki Male (7) SD	Suzuki Female (48) Mean	Suzuki Female (48) SD	Rhine et al. (1982) American Whites Male (37) Mean	Rhine et al. American Whites Female (19) Mean	Rhine and Campbell (1980) American Blacks Male (44) Mean	Rhine and Campbell American Blacks Female (15) Mean	Rhine (1983) South-western Indians Male (9) Mean	Rhine South-western Indians Female (2) Mean
Vertex	3.8	3.5								4.3	3.5	4.8	4.5	5.0	4.3
Forehead	3.9	3.6	4.2	3.6	3.9	3.0	0.24	2.0	0.14	5.3	4.8	6.3	6.3	5.8	4.5
Glabella	4.1	3.4	5.5	4.1	5.4	3.8	0.29	3.2	0.21	6.5	5.5	6.0	5.8	6.8	5.0
Nasion	5.3	4.6	6.6	3.0	4.8	4.1	0.26	3.4	0.31						
Nasal bone	3.1	2.8	5.4	2.5	3.8	3.9	0.22	3.4	0.28	3.0	2.8	3.8	3.8	3.5	3.3
End of nasals	2.2	2.1	2.4	2.9	3.4	2.2	0.19	1.6	0.14						
Subnasale	11.4	9.9	11.2	9.6	12.2	11.6	0.86	9.4	0.56	10.0	8.5	12.3	11.3	9.3	8.5
Midphiltrum	9.8	8.2	11.7	9.8	13.6					9.8	9.0	14.0	13.0	9.8	10.0
Upper lip										11.0	10.0	15.0	15.5	11.0	11.3
Lower lip										10.8	9.5	12.0	12.0	11.3	11.0
Labiomental	9.9	10.4	11.0	9.2	10.5	10.5	0.39	8.5	0.71	11.6	10.0	12.3	12.3	12.0	13.3
Mental	10.0	10.1	11.0	9.1	9.8	6.2	0.66	5.3	0.52	7.3	5.8	8.0	7.8	8.0	7.8
Menton	4.8	3.2	6.2	5.7	5.3	4.8	0.74	2.8	0.36	4.3	3.5	8.5	8.0	4.3	4.3
Lateral forehead										4.3	3.5	8.5	8.0	4.3	4.3
Supraorbital	5.7	5.3	6.6	5.1	6.9	4.5	0.33	3.6	0.32	8.3	7.0	4.8	4.5	9.0	8.3
Infraorbital	4.2	4.5	5.5	5.2	5.7	3.7	0.39	3.0	0.20	5.8	6.0	7.7	8.3	7.5	6.8
Lateral orbit						2.6	0.21	2.9	0.28	10.0	10.8	13.1	13.5	12.5	13.8
Zygomatic attach	7.71	7.7	10.0	4.9	7.3	5.4	0.46	4.7	0.66						
Canine fossa										13.3	12.8	16.7	17.5	14.0	15.0
Upper 1st molar						14.5	0.98	12.3	1.15	19.5	19.3	22.1	21.0	21.5	19.0
Lower 1st molar						10.2	0.59	9.7	0.72	16.0	15.5	16.2	17.0	19.3	15.8
Midmandible	8.2	7.1	7.1	10.1	9.7										
Zygomatic arch	4.5	4.1	5.8	8.1	4.5	4.4	0.25	2.9	0.28	7.3	7.5	8.7	9.3	7.5	
Midmasseter	17.1	15.9	20.1	20.5	18.6	13.6	0.63	10.4	0.86	18.3	17.0	19.3	18.8	20.8	
Gonion	6.6	3.6	11.7	17.5	13.6	6.8	0.78	4.9	0.72	11.5	12.0	14.5	14.3	13.3	
Supraglenoid	8.5	7.1	8.6	7.4	11.0					8.5	8.0	11.8	12.2	8.5	
Opisthocranion	4.0	3.1													

[a] combined results from table 5.2

Note: SD = standard deviation.

Table 5.4 Comparison of facial tissues between unembalmed and embalmed Caucasian cadavers.

Facial points	Male (17) Mean	SD	Female (23) Mean	SD	Unembalmed (9) Mean	Recently embalmed (9) Mean	Fully embalmed (31) Mean
Forehead	5.5[a]	1.88	4	1.43	4.22	6.56	4.65
Superciliary	8.17	2.37	6.82	1.93	5.92	8.03	7.39
Glabella	6.69	1.77	5.83	1.37	5.11	7.67	6.19
Nasion	6.69*	1.41	5.32	1.19	4.89	7.33	5.92
Rhinion	3.04	1.03	2.59	0.99	2.22	3.22	2.78
Zygomatic arch	10.88	4.9	9.07	2.83	7.53	12.64	9.83
Maxillary	17.42	3.68	15.64	4.34	11.75	19.39	16.39
Alare	11.44	3.36	11.4	2.59	8.28	13.31	11.42
Supracanine	8.81	2.68	7.56	2.32	6.03	10.31	8.08
Subnasale	13.46*	2.97	10.89	3.27	8.94	14.61	11.97
Philtrum	10.15	3.29	8.31	2.54	5.94	9.78	9.08
Upper lip	8.58*	2.63	6.78	1.94	5.39	9.5	7.53
Lower lip	9.62*	2.23	7.58	2.05	7.06	11.5	8.44
Labiomental	11.08	2.47	9.79	2.44	8.78	12.94	10.35
Pogonion	8.04	2.71	8.89	2.71	9.28	12.28	8.53
Gnathion	7.36	2.68	6.89	2.2	6.56	9.61	7.1
Gonion	18.52	10.6	13.61	5.22	11.56	23.17	15.71
Midmandible	12.21	4.85	12.13	5.11	8.67	16.67	12.17
Mandible border	12.47	7.13	9.87	3.73	6.81	12.97	10.84
Midmasseter	21.04*	4.73	17.6	3.73	16.67	26.03	18.98

[a] Significantly thicker by sex.
Source: Modified from Simpson and Henneberg, 2002.

the face, especially in older subjects. Special conditions must be followed regarding radiation levels and movement of the subject, and archival scans may not be studied as the required views are not always available. One further problem arising from these studies is that the subjects are exposed to radiation and a wide sample is not always possible to achieve. However, many studies have been carried out into tissue depth measurements using these imaging techniques (Bankowski, 1958; Weining, 1958; Leopold, 1968; George, 1987; Phillips & Smuts, 1996; Sahni, 2002) and the results are considered to be more accurate than those of the previous cadaver studies (see Tables 5.5 and 5.6). Using the craniographs of White Americans, Weining (1958) measured 99 males and 21 females, aged 17–42 years, and George (1987) studied 17 males and 37 females, aged 14–34 years. Other radiographic research studied White Europeans. Bankowski (1958) studied 15 males and 9 females, aged 20–84 years, and Leopold (1968) measured 102 males and 52 females, aged 16–82 years. Another radiographic study (Auslebrook *et al.*, 1996) measured the faces of 55 male Zulus (see Table 5.6). Phillips and Smuts (1996) studied 32 Mixed Race subjects (16 male and 16 female), aged 12–71 years, using CT scanning, and Sahni (2002) measured the facial tissue depths of 60 Indians (30 male, 30 female), aged 18–40 years, using MRI scans (see Table 5.6). When the combined White European cadaver studies are contrasted with the live subject studies (Bankowski, 1958; Weining, 1958; Leopold, 1968) it can be seen that tissues are significantly smaller on the cadaver at the majority of points. The exception is at the rhinion (end of the nasal bones), where the tissues are very thin and, presumably, less affected by postmortem changes.

Ultrasonics as an aid to measuring the human body was developed around 1960. Ultrasound has been used for many years in different applications: to detect flaws in metal, to detect the thickness of tissue in livestock and in biological applications (Whittingham, 1962). Ultrasound is emitted as a beam, and in medical practice the useful frequency range is generally 1–6 MHz. Ultrasonic measuring techniques are considered to be safe and effective. Baker and Dalrymple (1978) suggest that '*there is apparently*

Table 5.5 Facial tissue measurements using medical imaging techniques.

| | Weining (1958) Craniographs White Americans | | | | Bankowski (1958) Craniographs White Europeans | | | | Leopold (1968) Craniographs White Europeans | | | | George (1987) Lateral craniographs American Whites | |
| | Male (99) | | Female (21) | | Male (15) | | Female (9) | | Male (102) | | Female (52) | | Male (17) | Female (37) |
Facial points	Mean	Range	Mean	Range	Mean	Range	Mean	Range	Mean	Range	Mean	Range	Mean	Mean
Vertex	6.6	5–9.2	5.8	4–8.3	5.2	2.5–8.2	5.5	2.4–7.7	5.6	2.5–9.5	5.5	2.5–8		
Forehead	5.7	3.3–8.1	5.4	4.2–6.6	4.8	2.9–6.7	4.1	1.9–5.9					5	4.5
Glabella	6.3	5.1–9.2	6.3	4.6–7.6	5.6	3.8–7.7	5.8	2.9–7.6	6.2	3–10	6.2	3.5–9	7	6
Nasion	9	4.3–12.9	8.3	6.7–10.2	8.2	5.7–9.7	7.2	4.3–8.7	8.2	4.9–11	7.2	3.5–10.5	8	7.5
Midnasal bone									5.2	3–8	4.5	2–7		
Rhinion	3.1	1.7–5.3	2.6	2–4.3	2.8	1.9–4.3	2.3	1.7–2.9	3.3	2–5.5	2.8	1.5–5	3.5	3
Subnasale	14.8	8.5–22.6	12	9–16	16.4	12.5–20	13.4	9.1–17.3	15.4	9–23	12.4	6–18	3	
Midphiltrum	15.2	12.1–21.1	13.2	10.1–17.6	11.7	8.6–15.3	9.1	6.7–10.5					17.5	14.5
Upper lip									13.4	8–18.5	10.6	8.2–18.8	15	12.5
Lower lip									15.9	9–25	12.8	8–21	17.5	14.5
Labiomental	13	9.7–15.7	11.7	9–14.1	11.6	9.6–14.4	11.7	9.7–13.4	12.5	7–16.5	11.8	7–18.5	12.5	12
Mental	13.3	7.6–18.3	12.7	9.2–18.1	13.6	11.5–16.7	12.2	10.5–14	14.8	7.5–19	13.3	6.5–19	13	11.5
Menton	8.9	5.7–13	8	6–9.6	9.6	7.2–12	7.8	5.7–9.7	10.5	7–19	8.6	4–14.5	10.5	8.5
Lateral forehead														
Supraorbital														
Infraorbital														
Lateral orbit														
Upper molar														
Lower molar														
Zygomatic arch					5.7	3.3–8.8	4.8	2.7–7.7						
Midmasseter														
Gonion					8.6	4.3–17.4	7.8	5.2–14.5	10.8	6–17	10.3	5–17		
Supraglenoid														
Opisthocranion					6.6	4.3–9.6	5.7	2.4–7.7	8.6	5–14.5	8.1	4–12		

Table 5.6 Facial tissue measurements using medical imaging techniques.

	Phillips and Smuts (1996) CT scans Mixed Race South Africans				Sahni (2002) MRI scans Indians				Auslebrook et al. (1996) Radiographs Zulus		Ultra sound Zulus	
	Male (16)		Female (16)		Male (30)		Female (30)		Male (55)		Male (55)	
Facial points	Mean	SD	Mean	SD	Mean	SD	Mean	SD	Mean	SD	Mean	SD
Forehead	5.36	1.44	4.88	1.02	3.6[a]	0.14	3.4	0.15	5.21	0.92	5.53	0.88
Glabella	5.47	0.68	5.64	1.42	4.7[a]	0.15	4.2	0.1	5.76	0.88		
Nasion	4	2.42	4.68	2.35	5.1[a]	0.12	4.6	0.14	7.03	1.11		
Midnasal									4.82	1.04		
Rhinion	2.88	1.08	2.78	0.91	2.1	0.1	2.1	0.16	3.08	0.58		
Subnasale									12.8	2.44		
Midphiltrum	12.25	2.97	10.13	2.48	11.4[a]	0.2	9.2	0.33	12.1	1.63	9.79	1.68
Upper lip	13.16	2.51	13.63	3.7	9.6[a]	0.1	8.6	0.18	14.61	2.17	9.52	1.44
Lower lip	10.43	1.69	12.45	2.31	12.5	0.15	12.8[a]	0.2	16.38	1.96	10.3	1.42
Labiomental	12.02	2.07	11.7	1.66	9.6[a]	0.13	9.5	0.14	12.87	1.65		
Mental	8.94	2.42	9.57	2.36	9.7[a]	0.1	9.6	0.14	11.66	1.79	8.99	1.87
Menton	6.61	1.71	6.47	1.57	6.4	0.12	6.6[a]	0.12	7.26	1.98		
Lateral forehead	4.51	1.4	4.78	1.74	4.5[a]	0.15	4.1	0.15	5.37	1.01	4.79	0.72
Supraorbital	5.46	1.31	5.79	1.89	5.8[a]	0.11	5.6	0.11	6.18	0.98	6.05	0.87
Infraorbital	5.97	2.87	6.42	3.83	4.5	0.2	4.7[a]	0.22			6.56	1.88
Lateral orbit	7.54	1.49	8.25	2.52	6.2[a]	0.26	5.9	0.24				
Upper molar	12.68	2.1	12.99	4.45	18.9	0.46	18.8	0.44				
Lower molar	13.13	5.31	11.88	5.95								
Zygomatic attach									7.97	1.77	7.02	1.05
Alare									8.18	3.28	9.47	2.16
Supracanine									11.73	1.4	15.38	2.6
Subcanine									12.08	1.59		
Midmandible											10.42	1.63
Zygomatic arch	6.49	2.5	9.3	3.21	5.4[a]	0.22	4.4	0.23				
Midmasseter	19.06	9.08	21.26	8.37								
Gonion	14.2	6.08	13.5	6.6	10	0.25	9.85	0.24			18.05	1.69
Supraglenoid	9.1	4.04	8.44	3.84	8.9[a]	0.19	7.55	0.24				
Opisthocranion					6.4	0.24	6.3	0.22			5.91	1.35

[a] Significantly thicker by sex.

little or no danger associated with diagnostic ultrasound exposure at clinical levels'. Many studies into facial tissue measurements have been carried out using this ultrasonic technique, and ultrasound is considered to be the most accurate method. One study (Booth *et al.*, 1966) compared ultrasonic, conductivity and caliper subcutaneous fat measurement techniques on 41 subjects at the abdomen and infrascapular sites. They found that the ultrasonic technique was the most reliable, with imperceptible differences between measurements at the same site. Another study (Bullen *et al.*, 1965) compared ultrasonic and needle-puncture subcutaneous fat-plus-skin measurement techniques on 100 subjects at the abdomen and triceps sites. They found a high reliability for the ultrasound technique and a high correlation between the needle-puncture and ultrasound techniques. In addition, Stouffer (1963) refers to a study in which direct comparison was made between ultrasound, x-ray, ruler probe and lean meter measurements, in which ultrasound was found to be comparable to x-rays in accuracy and more accurate than the other techniques. In addition, a study (Lauprecht *et al.*, 1957) was carried out comparing ultrasonic measurements with direct measurements of tissue thickness (with a ruler) from slaughtered animals, and the ultrasonic measurements were in agreement with the direct measurements. Ultrasonic studies using human tissue have been carried out (Whittingham, 1962) involving the examination of six patients before operation and measuring after incision by ruler, at points 4 cm either side of the umbilicus. The ultrasonic measurements were in good agreement with the ruler measurements.

One problem associated with the ultrasonic technique is the accuracy of placing the pen at the correct anatomical point over the skull. These points have to be found using surface anatomy and palpation, and there has been some documented evidence of errors. Suk (1935) conducted experiments indicating that superficial palpation does not accurately locate underlying landmarks. His results showed that the nasion landmark was missed by an average of 3 mm. Another study by Farkas (1994) stated that the determination of some points is very difficult if the angle is flat or if there is rich soft tissue cover. Finally, Nelson and Michael (1998) suggested that the determination of bony points from surface landmarks is subjective and inaccurate, and he stated that the most difficult points are midmasseter, chin–lip fold and mid-supraorbital. However, the vast majority of facial tissue depth measurements can be performed with a positional accuracy appropriate to facial reconstruction requirements. There have been suggestions (Nelson & Michael 1998) that the angle of the pen to the bone will also affect the measurement, but since the pen will only record waves that bounce back from the bone along the path perpendicular to the surface of the bone, this does not seem likely. When the European ultrasound studies (Helmer, 1984) are contrasted with the Victorian cadaver studies (His, 1895; Kollman & Buchly, 1898) it can be seen that all the points on the cadavers show smaller measurements (see Tables 5.2 and 5.7). When the American ultrasonic tissue measurements (Manhein *et al.*, 2000) are compared with the later American cadaver measurements (Rhine & Campbell, 1980; Rhine *et al.*, 1982), the results are much more similar (see Tables 5.3 and 5.8). There are still thinner cadaver tissues in the White American results at the cheek and gonion areas, and thinner cadaver tissues in the Black American results at the rhinion (end of nasal bone), philtrum, cheek and gonion areas. Rhine and his colleagues were careful not to include cadavers with distortion caused by postmortem changes, and this may have rendered their results more comparable with live subjects. Many other facial tissue studies have been carried out using ultrasonic measurement techniques. Auslebrook *et al.* (1996) studied 55 Zulu men aged 20 to 35 years, and employed ultrasonic readings, in addition to cephalometric radiograph measurements, to create average sets of facial tissue depths (see Table 5.6). Helmer (1984) studied 123 (61 male and 62 female) White European subjects between the ages of 20 and 80 years. The results were divided into ten-year age groups and by sex, and provide a comprehensive and accurate database (see Table 5.7). Finally, Lebedinskaya and her colleagues (1993) studied nine ethnic groups within the Soviet Union using ultrasonic echolocation. She studied Koreans, Buryats, Bashkirs, Abkhazians, Russians, Lithuanians, Armenians, Uzbeks and Kazakhs to a total of 1695 individuals between the ages of 20 and 50 years (see Table 5.9).

Table 5.7 Adult White European facial tissue measurements.

Facial points	20–29 years Male (13) Mean	Range	20–29 years Female (12) Mean	Range	30–39 years Male (14) Mean	Range	30–39 years Female (13) Mean	Range	40–49 years Male (13) Mean	Range	40–49 years Female (11) Mean	Range	50–59 years Male (11) Mean	Range	50–59 years Female (15) Mean	Range	60+ years Male (10) Mean	Range	60+ years Female (11) Mean	Range
Vertex	5	4–6.5	4.5	4–5.5	5	4–6	5	4.5–5.7	5	3.5–7.5	5	4–5.5	5	4–65	5	4–7	4.8	3.5–6.7	5	3.5–6.3
Trichion	4.3	3.5–5.5	4.1	3.5–5	4.7	3.3–6	4	3.5–5	4.5	3.5–5.7	3.9	3.3–5.2	4.7	3.7–6.3	4	3–4.7	4.9	3.8–5.8	4	3.7–5.5
Metopion	5	4–5.5	4.5	3.5–5.2	5	3.7–6.5	4.5	4–5.7	5	3.2–6.5	4.6	3.5–5.2	5	4–6	4.7	3.5–5.2	4.8	4–6	5.2	4.2–5.5
Ophryon	5.5	5–6 – 7	5	4.2–5.8	5.8	4.5–7.5	5.2	4.5–5.8	5.5	3.8–7	5.3	3.8–6	5.8	5–6.5	5.3	3.8–6.3	5.8	4.5–7.3	5.8	5–7.2
Glabella	5.7	5.2–6.7	5.5	4.5–6.3	6.2	5–7.7	5.7	5–6.5	6	4.3–7.5	5.9	4.5–6.7	6	5.5–7	6	4.5–7	6.3	5–7.7	6.5	5.3–7.3
Nasion	8.2	6.3–10.2	6.9	4.7–7.3	7.3	6–11.3	6.5	6–77	6.8	5.3–8.3	6.2	5.3–7.5	7.3	6.2–8.2	6.5	4.8–7.5	7.1	6–9.2	6.5	5.5–7.7
Nasal bone	3	1.5–4.5	2.9	1.7–4.0	3.5	2–4.2	3	1.5–4.5	3.9	2.5–4.5	3.5	1.5–6.3	3.5	1.5–6.3	3	1.5–4	3.7	2.8–5	3	2.2–3.8
End of nasal	2.3	1–3.7	2.3	1.5–4.2	2.5	1.5–4.0	2.5	1–3.8	2.7	1.8–4.5	2.4	1–3.2	2.8	1–4	2.3	1.5–4.3	2.6	1.3–4.3	2.5	1.8–3.5
Lateral nasal	7.5	5.8–9	7	4.8–8.5	7.4	5.2–9.5	6.3	4.7–8	7.3	6–9.5	6.7	4.5–9.8	8.2	6.3–12.8	6.5	5.5–9.7	6.7	5.7–10	7.3	5–10.2
Alare	13.3	11.2–14.5	11.6	9.3–13	11.7	10–16.2	11	9.2–13.7	12.2	9.3–14.7	11	9.2–12.3	12.5	10.8–13.7	11.5	10–12.8	11.9	9.7–14.3	11.5	9.5–12.8
Subnasale	15.5	12.3–17.2	13.8	11.5–15	14.6	12.5–18.7	12.8	11.5–14.5	15.6	9.5–19	12.6	11.5–14.5	14.3	12.2–19.5	13.2	10–15.7	12.9	8.5–15.3	12.2	9.7–14.2
Upper lip	14	11.8–17	11.8	10.2–14.2	12.3	9.5–17.3	10.7	9–12.8	12.6	9–18.2	10.5	8.7–11.7	11.8	9.3–14.8	10	8.2–14.5	9.9	7.7–11.7	9.8	8.7–10.3
Lower lip	14.2	10.5–16	12	10.8–15.7	14.9	12.3–17.2	12	10–13.7	14.2	11.5–19.2	12.5	9.5–14	13	11.5–16.8	11.8	9.8–16.2	12.7	11.5–16	11.5	9.7–15.3
Labiomental	12	10–14	10.4	9.3–12.5	12.1	10.7–14.3	10.8	8.8–12.7	13.3	10–15.8	12.3	9.5–14.2	13	12–16.3	12.2	10.2–13.2	12.8	11.2–15.3	12.7	10.3–15.3
Pogonion	9.7	7.3–13.7	9.6	6.7–11.3	10.3	7.8–13	10	7.3–12.2	11.7	8.3–18.2	9.6	5.5–11.8	13.7	9.7–17.3	11.3	8–13.5	12.3	10.3–14.7	12	8.8–13.8
Gnathion	7.5	6.5–10	7.1	5.6–8.7	8.3	6–9.7	7.2	5.3–8.8	9.5	6.5–13.7	6.9	5–11	9.8	7–12.7	8	5.5–11.3	8.9	8–11.7	8.7	6–12.2
Lateral forehead	5.5	4.5–6.7	5.2	4.5–5.7	6	5–7	5	4.5–5.7	5.5	4.2–7.3	5.3	4–6.3	6	4.5–8	5	4–6.3	6.2	4.8–7.5	5.3	4.5–6.5
Mid-supraorbital	7.3	6.3–9.2	6.6	5.7–8	7.3	6.0–8.5	6.5	5.8–7.5	7.2	5.3–11.7	7.4	5.3–8.3	7.5	6.2–8.5	6.7	6–7.5	6.7	5.8–8.2	6.8	6.3–8
Orbitale	5.2	4.2–5.7	5.5	2.7–6.7	5	4.2–8.2	5.5	3.7–8.5	5.8	3.3–7.5	5.4	3.8–7.3	5.5	4.7–9.7	6	3.5–8	5.8	3.7–11.7	6.3	5.2–10.7
Canine fossa	18.8	16–23.7	18.8	15.7–21	19.7	16.3–24.7	20.2	16–27.7	21.5	18.2–25.7	19.1	15–23.3	21.8	17–24.5	20.7	14.7–24.7	21.5	15.7–24	22.3	18.2–28
Upper 1st molar	20.2	15.7–25.3	19.2	15.8–11.7	22	16–29	21.5	10–29.3	21.7	17.8–28.7	20.5	16.8–24	22.3	14–28.3	19.3	14.3–23.3	18.8	13.8–23.2	20.5	11.8–26.3
Lower 1st molar	19	14.5–24.3	16.6	13.8–20.8	18.5	15.3–23	19	14.8–29	18.3	16–24.3	18	15–22.8	18.3	16.3–21.8	17.7	15–21.7	17.2	14.7–19.8	19	11.3–24
Mandibular	9.2	6.2–11.8	9.2	6–11.5	10.1	5.2–15	9	5.3–13	10.2	6.3–14.8	9.1	7.5–12.5	12	8.8–15.7	12	5.6–11.3	10.3	9.5–12.5	10.3	8–13.7
Frontotemporale	5	4.3–6.0	5	4.3–5.5	5.3	4–6.7	5	4.3–8.7	5.5	3.8–7	4.8	3.5–6.5	5.5	4–7.8	5	3.5–6.3	5.5	4–6.7	5	4–5.3
Lateral orbit	5.3	4.3–5.7	5.2	4.2–6	5.2	4.3–6.2	5	4.3–5	5.8	5–7	5.1	3.5–7.5	5.7	4.8–7	5.3	4–5.8	5.6	4.2–6.7	5.5	4.5–7.2
Lateral zygomatic	7.5	6.2–9	8.9	7.5–9.8	7.6	6–9.5	9	7.2–10.7	6.8	4.8–9.3	9.1	6.5–11.3	8	6.3–12	9	6.2–10.8	7.5	5.3–9.5	10.3	6.7–13
Zygomaxillare	9.5	8.3–10.7	10.3	8.8–12.3	9.9	5.5–12	10.3	8.3–12.8	10.1	7.5–12.2	10.6	7.8–14.3	10.7	9.2–13.2	10.7	6.5–12.7	9	6.5–11.2	12	7.5–15
Midmandible	12	10.7–14.3	10.7	8.0–13	11.9	6.5–16.7	11.5	9.7–15.7	12.8	8–19	11.8	6.8–16.7	14.2	11.8–20	10.7	9.8–15.5	13.4	11.2–16.3	13.7	11.3–19.3
Euryon	6	5–9.5	5	4.5–7	6.7	5–9	5.5	4–67	6.5	44.5–8	5	4.5–6.5	7	5.5–8	5.3	4–7.2	6.4	5.3–7.8	5.2	4.5–6
Temporalis	15.3	13.2–18	14.2	12–17	16.3	13.7–19.3	14.2	8.7–19.8	16.1	13–19.3	14.3	11–16.2	14.7	14.3–22.3	13.3	4.3–19.2	14.9	8.7–17.2	13.3	11.3–18.5
Zygomatic arch	5.3	3.5–6.3	4.8	4.3–5.8	5.3	4–9.7	5.2	4–8.5	5.5	4–8.5	5.4	4.3–7.8	5.5	4.8–10.8	5.3	4.3–7	5	3.8–8.2	5.2	4.5–10
Midmasseter	19.2	15–23.3	17.2	13.8–21	21.3	15.5–23	18.3	14.8–22.3	20.4	13.7–29.8	17.8	15.3–22.2	20.5	17.7–23	17.3	13.7–22	20.6	16.3–25.3	19.2	15.8–23.7
Gonion	11	6.3–15.8	11.6	9.2–16	13.2	8.5–16.3	11.7	9.7–16.2	13.3	3.5–1.8	11.2	9.5–14.5	11.7	6.7–17.7	10.3	4.3–14.5	14.4	7.5–17.3	14.3	11.2–17.3
Opisthocranium	5.5	4.5–6.5	4.5	4–6	5.5	4.5–6.5	5	3.5–5.5	5.5	4.5–7.5	5	3.5–6.5	5.5	4.5–7.5	5	4–7	5.5	4.5–7.7	5	3.5–6.3

Source: Modified from Helmer, 1984.

Table 5.8 Adult White and Black American facial tissue measurements.

	Black Americans										White Americans															
	19–34 years				35–45 years				45–55 years		19–34 years				35–45 years				45–55 years				>56 years			
	Male (91)		Female (91)		Male (95)		Female (167)		Female (99)		Male (91)		Female (91)		Male (95)		Female (167)		Male (84)		Female (99)		Male (155)		Female (0)	
Facial points	Mean	SD	Mean	SD	Mean	SD	Mean	SD	Mean	SD	Mean	SD	Mean	SD	Mean	SD	Mean	SD	Mean	SD	Mean	SD	Mean	SD	Mean	SD
Glabella	5.2[a]	1.12	4.6	0.7	5.3[a]	1.53	4.5	0.93	4.8	0.84	5[a]	0.67	4.8	0.95	5.5[a]	1.27	4.7	1.03	6[a]	1.41	4.8	1.17	5.6[a]	1.52	5.2	0.97
Nasion	6.6[a]	0.84	6	0.91	5.7[a]	2.08	5.2	1.25	6	1	6[a]	1.12	5.5	1.16	6.4[a]	1.43	5.3	1.39	7.2[a]	1.64	6.2	0.75	6.6[a]	1.52	6	1.22
Rhinion	2.2[a]	0.42	1.7	0.46	1.7	0.58	1.5	0.51	2	0.71	1.9	0.45	1.8	0.63	2.4	0.97	1.6	0.51	1.8	0.45	1.8	0.41	2	0	1.8	0.67
Alare	9.2	2.82	8.4	1.98	10.3	2.52	8.4	2.01	8.4	1.52	7.5	1.9	8.6	1.99	9.8	1.81	8	1.73	10.4	2.51	10.8	1.94	10.8	3.03	9.8	2.22
Midphiltrum	13[a]	2.2	9.2	1.82	11[a]	1.73	8.8	1.92	8.2	2.49	11.9[a]	2.24	9.1	1.69	10.6[a]	1.43	7.4	1.3	8[a]	3	8	1.41	9.4[a]	1.52	8	2.65
Labiomental	12.7[a]	2.05	11.8	2.2	12.7[a]	1.15	11.7	2.42	10	2.55	11.1[a]	1.85	10.3	1.55	13.1[a]	1.52	9.6	1.5	11.6[a]	1.67	9.8	2.32	12.2[a]	1.79	11.4	1.42
Mental eminence	12.1	2.9	10.8	2.68	12.3	4.51	11.2	2.25	10.8	3.11	10[a]	2.77	9.2	2.08	12[a]	3.2	9.2	2.14	11[a]	1.73	10.7	2.8	11.8	2.05	12.3[a]	1.58
Menton	8.8[a]	1.89	6.7	2.02	7[a]	2	6.4	2.65	7.2	1.92	7.2[a]	1.73	6	1.45	8[a]	1.05	5.4	1.84	7.2[a]	1.79	6.7	2.94	5.6	0.89	8[a]	1.87
Supraorbital	6.4	1.3	6.1	0.83	6.3	0.58	6	1.22	5.8	0.84	5.3	1.25	5.7	1.04	5.9	0.88	5.5	1.19	7.7	1.67	6.5	0.84	5.6	1.14	6.3	1
Infraorbital	5.8	1.26	6.2	1.17	7	1	6.9	1.96	5.8	1.3	5.8	1.58	6.1	1.05	6.2	1.87	5.7	1.33	6.8	0.84	7.3	4.08	5	2	7[a]	2.5
Supracanine	12.8[a]	1.86	10	2.28	10.3[a]	1.53	9.6	2.75	9	2.45	11.9[a]	2.65	9.3	1.74	10.1[a]	2.13	7.8	1.37	10[a]	2	7.7	1.86	9.2	1.1	8	2
Subcanine	14.4	2.89	10.9	2.44	10.7	0.58	11.5[a]	1.6	12.4	3.91	11.5[a]	2.17	9.4	1.56	10.2[a]	1.32	8.7	2.23	10[a]	2.35	9	2.97	11.8[a]	2.39	9.7	3.39
Posterior maxilla	28.2[a]	3.46	26.6	4.36	27.3[a]	4.51	26.8	4.47	26.8	4.09	28.5[a]	4.69	26.3	4.94	24.6[a]	6.45	25.1	6.74	28.2[a]	7.53	27.2	6.11	23.6	8.11	29.4	4.82
Upper molar	24.5	4.05	21.7	3.99	23.7	4.04	22.5	3.93	21.2	5.89	25.1	4.15	23.4	4.53	21.1	6.69	20.1	5.15	21.4	3.85	21.7	5.32	20.6	6.11	27.2	5.59
Lower molar	14.1	4.21	12.6	2.85	13.3	2.31	13.1	4.17	13.4	4.04	14.8	4.48	13.7	3.25	15.6	4.81	12.6	4.21	15.4	4.39	13	4.29	20.6	6.11	17.4	3.28
Lateral orbit	4.8	0.76	5	0.84	3.7	0.58	4.9	1.18	4.8	0.84	4.2	0.79	4.7	0.88	4.3	0.82	4.3	0.9	5.4	0.55	4.5	1.87	5.2	0.45	4.9	1.76
Zygomatic attach	8.4	2.22	10.2[a]	2.28	6.3	0.58	9.8[a]	2.38	9.8	3.27	7.8	2.38	9.3[a]	1.7	8.2	2.2	8.7[a]	2.74	5.4	2.05	10.2[a]	1.6	6.4	1.34	4.9	2.45
Gonion	21.1[a]	3.24	17	4.23	20.7[a]	2.89	16.2	3.64	14.8	2.86	20[a]	4.27	17.4	3.7	19.6[a]	5.87	15.3	4.5	19[a]	4.69	14.7	4.68	14	4.95	16.9[a]	3.59
Supraglenoid	7.4	1.77	6.4	2.25	5.7	1.15	5.6	2.22	6	2.24	7.8	2.29	7.4	2.07	6.6	3.86	4.9	1.44	5.4	1.52	6	1.55	5.2	1.1	7.4	2.3

[a] Significantly thicker by sex.

Source: Modified from Manhein *et al.*, 2000.

Table 5.9 Adult facial tissue measurements from different ethnic groups.

Facial points	Koreans Male (91) Mean	SD	Koreans Female (91) Mean	SD	Buryats Male (95) Mean	SD	Buryats Female (167) Mean	SD	Kazakhs Male (84) Mean	SD	Kazakhs Female (99) Mean	SD	Bashkirs Male (155) Mean	SD	Uzbeks Male (55) Mean	SD	Uzbeks Female (71) Mean	SD
Forehead	4.5	0.98	4.5	0.89	4.5	0.88	4.7	0.95	4.5	0.87	4.9	0.9	5.1	0.85	5.1	0.71	5	0.71
Superciliary	5.2	0.81	5.2	0.86	5.4	0.79	5.7	1	5.2	0.82	5.6	0.87	5.6	0.89	5.4	0.76	5.5	0.77
Glabella	5.1	0.8	5.4	0.89	5.4	0.75	5.6	0.88	5.3	0.79	5.6	0.86	5.6	0.84	5.4	0.75	5.5	0.77
Nasion	4.5	0.79	4.4	0.86	4.8	0.85	4.5	0.89	4.8	0.91	4.6	0.7	5.8	0.85	5.7	0.87	5.3	0.77
End of nasal	2.8	0.31	2.9	0.35	2.8	0.43	2.8	0.3	3	0.38	2.9	0.38	3.8	0.56	4.1	0.68	4	0.56
Lateral nasal	2.9	0.31	2.9	0.28	2.9	0.33	2.9	0.33	3	0.36	3	0.33	4	0.75	3.9	0.66	3.9	0.58
Maxillary	13.2	1.86	13.9	1.65	14.5	1.96	15.8	1.79	13.2	1.63	14.5	1.9	11.6	2.36	14.1	1.88	15.5	2.14
Zygomatic attach	9.8	1.85	12.2	2.02	10.6	1.77	13.6	1.78	9.8	2.02	12.6	2.09	9.3	1.47	9.3	2.04	11.7	1.93
Zygomatic arch	4.7	0.8	5.6	0.9	4.5	0.89	5	0.77	4.5	0.78	5.3	0.88	5	0.93	4.5	0.58	5	0.7
Supracanine	10.4	1.33	9.3	0.95	10.8	1.21	9.8	1.04	10.7	1.34	9.9	1.01	10.1	1.34	10.2	1.66	9.8	1.04
Philtrum	11.1	1.44	9.6	1.13	11.8	1.52	10.2	1.23	11.7	1.4	10.3	1.3	11.6	1.64	11.9	1.63	11	1.27
Upper lip	12.6	1.73	10.6	1.57	13.5	1.9	11.7	1.81	12.4	1.7	11.1	1.53	13	1.9	13.1	2.02	12.1	1.51
Lower lip	13.8	1.51	12.3	1.49	14.5	1.63	13.1	1.73	13.7	1.61	12.4	1.42	14.5	1.72	14	1.98	13.1	1.52
Labiomental	11.3	1.34	11.1	1.16	11.7	1.53	11.2	1.37	11.2	1.07	11.1	1.2	11.3	1.47	11.2	1.46	10.8	1.4
Pogonion	10.6	1.85	11.1	1.71	11.4	1.93	11.9	1.82	10.9	1.66	11.4	1.53	10.9	1.88	11.2	1.9	10.6	1.52
Gnathion	6.3	1.17	6.5	1.12	6.8	1.18	6.9	12.8	6.4	1.25	6.6	1.21			6.4	0.97	6.3	1
Lower 2nd molar	12.8	3.43	14.6	2.83	13.1	3.12	14.8	2.54	12.6	2.8	14.6	2.72	10.1	2.26	11.4	2.94	13.1	2.4
Midmandible	6.1	1.62	6.9	1.53	6.2	1.43	7.2	1.57	5.6	1.22	7	1.58			6	1.46	6.5	1.09
Midmasseter	17	2.26	17	2.18	17.2	2.02	17.5	1.67	17	2.06	16.9	2.13			16.8	2.02	16.9	1.95
Gonion	4.6	0.96	5.4	1.22	4.5	0.94	5.1	1.01	4.6	0.79	5.2	1.24	5.4	1.07	5.1	0.72	5.3	0.99

Table 5.9 (cont.)

	Armenians				Abkhazians				Russians				Lithuanians			
	Male (95)		Female (167)		Male (84)		Female (99)		Male (155)		Female (61)		Male (55)		Female (71)	
Facial points	Mean	SD	Mean	SD	Mean	SD	Mean	SD	Mean	SD	Mean	SD	Mean	SD	Mean	SD
Forehead	4.7	0.81	4.9	0.91	4.5	0.72	4.6	0.77	5.3	0.86	5.3	0.77	4.7	0.83	4.6	0.67
Superciliary	5.2	0.83	5.8	1.09	5.2	0.72	5.4	0.63	5.8	0.98	5.9	0.95	5.1	0.72	5.3	0.7
Glabella	5.3	0.9	5.7	0.98	5.2	0.74	5.4	0.75	5.8	0.79	6	0.89	5.5	0.75	5.5	0.78
Nasion	5.8	0.89	5.7	0.84	5.8	1.15	5.4	0.84	5.6	0.94	5.5	0.9	5.4	0.96	5	0.77
End of nasal	3.2	0.47	3.4	0.62	3	0.41	3	0.61	3.8	0.81	3.7	0.7	3.1	0.2	3.1	0.25
Lateral nasal	3.3	0.51	3.5	0.51					3.9	0.83	3.8	0.75	3.1	0.27	3.2	0.22
Maxillary	13.2	2.58	15.2	1.84					12.4	2.36	14.2	2.49	12.4	1.83	13.5	1.39
Zygomatic attach	9.3	1.31	12.3	2.09					9.8	1.6	12.4	1.97	9.3	1.64	11.7	1.77
Zygomatic arch	4.8	0.66	5.3	0.96					5.1	0.87	5.4	0.92	4.7	0.64	4.9	0.65
Supracanine	10.5	1.41	9.6	0.93	10.7	1.34	9.7	1.24	10.5	1.3	9.7	1.14	11.2	1.32	9.6	1.15
Philtrum	12	1.53	10.1	1.06			9.7	1.04	11.5	1.59	10.6	1.49	12.5	1.45	10.6	1.49
Upper lip	12.8	1.75	10.8	1.52	12	1.8	10	1.47	12.4	1.91	10.9	1.77	13.2	1.83	11	1.79
Lower lip	14.3	1.51	12.2	1.45	13.3	1.77	11.9	1.51	13.8	1.75	12.3	1.7	14.1	1.6	12.2	1.57
Labiomental	11.2	1.19	10.4	1.21	11.7	1.5	11.5	1.78	11.5	1.4	11.1	1.21	11.1	1.26	10.5	1.4
Pogonion	11.2	1.89	10.8	1.57	11.7	1.89	11.3	1.87	11.6	1.83	11.8	1.74	11.5	1.76	11.1	1.53
Gnathion	6.8	0.88	6.3	0.9									6.7	0.94	6.2	0.98
Lower 2nd molar	13.3	2.51	14.3	2.77					12	3.07	13.8	2.65	13.2	3.15	14.6	2.55
Midmandible	6.8	1.24	7.1	1.38									6	1.07	6	1.15
Midmasseter													18	2.08	17.5	2.1
Gonion	5.2	0.82	5.5	1					5.2	1.05	5.3	0.98	4.7	0.76	4.7	0.85

Modified from Lebedinskaya *et al.*, 1993.

Differences in facial tissues related to nutritional condition

In 1895, the anatomist, His, was the first to compare the facial tissue depths of emaciated and well-nourished people (see Table 5.1). He found that all the tissues were thinner for emaciated than well-nourished cadavers, except at the nasal bridge (midnasal bone), where the tissues were very thin for both groups. The most demonstrable differences were seen at the chin mental eminence, mandible and cheek regions, and under the chin area (menton), where some of the tissues were up to 4.5 mm thinner. Kollman and Buchly (1899) also studied the effects of poor nutrition on the tissues of the face. Their results were in agreement with His (see Table 5.1). They found that the majority of facial tissues were thinner in the emaciated group than the well-nourished group, except at the nasal bridge where there was no change. The most apparent variation was at the chin, cheek and jaw regions. Studies of White Americans (Rhine et al., 1982) and South-western Indians (Rhine, 1983) were in agreement with these conclusions. A later study of Japanese faces (Suzuki, 1948) found that, in general, the facial tissues decreased with a decrease in nutrition, but that the tissue around the eye did not decrease. He suggested that this may be due to the lack of subcutaneous fat at this area, and that the areas most affected were those with high levels of hypodermic fat or well-developed muscles, such as the mouth, cheeks and jaw. Sutton (1969) found that the tissues at the zygomatic arches, where there is substantial subcutaneous fat, were severely affected by the body-build of the individual. He found that the tissue depth for emaciated men was almost a third of that for fat men, and for emaciated women it was half of that for fat women.

Differences in facial tissues related to sex

The White European cadaver studies were in agreement regarding the differences in facial tissue depths between men and women (see Table 5.2). The combined results suggest that men have more tissue than females at the majority of facial points (see Table 5.3). The male and female tissues were similar at the nasion, nasal bones, end of nasal bone (rhinion), labiomental, mental, menton and infraorbital points. Women had more tissue at the zygomatic arch point only. This suggested that men had more tissue at the brow, mouth and jaw regions than women, and women had more tissue at the cheeks than men. All ethnic group studies (Von Eggeling, 1909; Suzuki, 1948; Rhine, 1983; Rhine & Campbell, 1980; Phillips & Smuts, 1996) showed sexually dimorphic results (see Table 5.3). Japanese female measurements were smaller than the male at the majority of areas (Suzuki, 1948). Japanese women had more tissue at the lateral orbit and zygomatic attachment points. This is in agreement with the White European studies. The Black American results (Rhine & Campbell, 1980) suggested that women had similar or slightly less tissue at all points, except at the cheeks (infraorbital, zygomatic arch and lateral orbit). This is in agreement with the White European and Japanese studies. The White American results (Rhine & Campbell, 1980) also suggested that women had similar or slightly less tissue at all points, except at the cheeks (infraorbital, zygomatic arch and lateral orbit). This is in agreement with the White European, Black American and Japanese studies (see Table 5.3). The South-western Indian study (Rhine, 1983) showed that women had similar or slightly less tissue at all points, except at the mouth and chin (upper lip, lower lip and mental). This was a different result to the other ethnic groups, but may be a reflection of the small number of females in this study (two). Simpson and Henneberg (2002) studied Australian White cadavers. They found that men had significantly more tissue at the forehead, mouth and jaw, and that the majority of measurements were greater for men than women (see Table 5.4). They found that women had greater measurements than men at the mental point only, but this was not a significant difference. Sutton (1969) also studied White Australian cadavers and found that women had more tissue at the zygomatic arch than men. These White Australian results are in agreement with the White European, Black American, White American and Japanese studies.

The imaging studies showed similar results. The radiographic studies of White Americans (Weining, 1958; George, 1987) suggested that men had more tissue at the majority of points than women, although similar tissue was found at the glabella, rhinion, labiomental and mental points (see Table 5.5). The White European studies (Bankowski, 1958; Leopold, 1968) also suggested that men had more tissue at the majority of points than women. Similar measurements for men and women were found at the vertex, glabella, rhinion, labiomental, mental and opisthocranion points. The CT scan study of Mixed Race South Africans (Phillips & Smuts, 1996) also found that both sexes had very similar tissue measurements (see Table 5.6). Although not significantly different, the measurements were greater in men than women at the forehead, jaw and chin points, and greater for women than men at the mouth and cheeks. The Indian results (Sahni, 2002), taken using MRI scans, suggested that all tissue was significantly thicker for men than women, except at the rhinion, opisthocranion, gonion and upper molar points, where the measurements were similar for men and women (see Table 5.6). Women had significantly thicker tissue than men beneath the chin and at the infraorbital and lower lip points.

The ultrasound studies also showed similar results. Helmer's study (1984) of White Europeans was divided into ten-year age groups and by sex, and provides a comprehensive database (see Table 5.7). The results showed that female tissues were thinner than male tissues at most areas of the face, at all ages (see Fig. 5.2). Similarity was seen at the vertex, nasal bone, rhinion, mental, infraorbital, canine fossa, lateral orbit, gonion and opisthocranion points. Exceptions were seen at the cheeks (zygomatic arch and zygomatic attachment) in females at all ages, and at the brow (forehead, glabella and supraorbital) of females over the age of 60 years, where the tissue was thicker than in males. Lebedinskaya and her colleagues (1993) found that female faces of nine different Russian ethnic groups showed similar, or slightly thinner, tissue than male faces, with the greatest exceptions in the region of the eyes and cheekbones, where the tissue was thicker for women than men (see Table 5.9). The study of White Americans (Manhein et al., 2000)

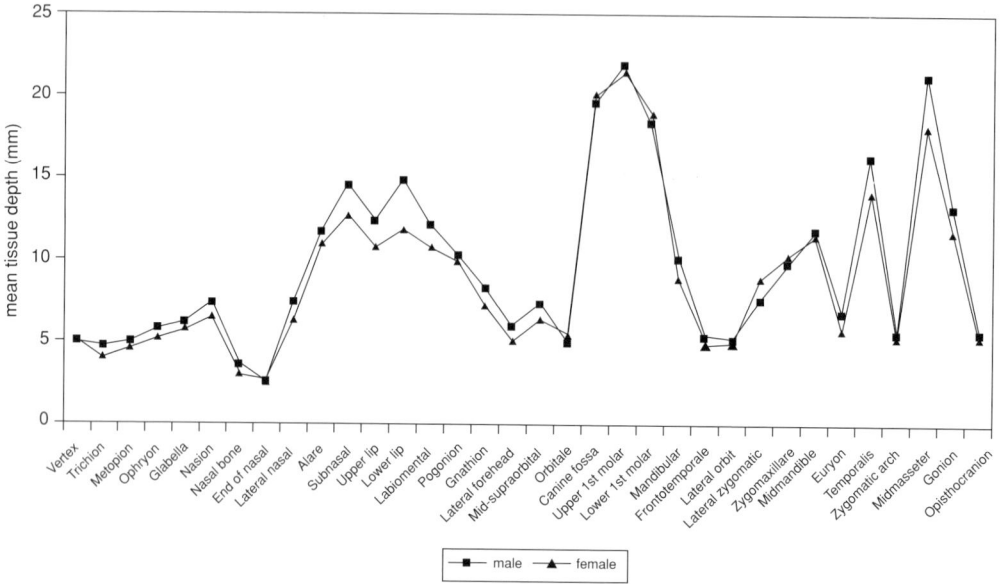

Fig. 5.2 Comparison of facial tissues between White European men and women. Modified from Helmer, 1984.

showed that men, of all ages, had significantly thicker tissue than women at the majority of points, except the rhinion, supraorbital, upper molar, lower cheek, lateral nostril, infraorbital, lateral orbit and supraglenoid points, which were similar for men and women (see Table 5.8). Women of all ages had significantly greater tissue than men at the zygomatic attachment, and women over 56 years had significantly more tissue at the chin, infraorbital and gonial points. The study of Black Americans (Manhein *et al.*, 2000) showed that men, of all ages, had significantly thicker tissue than women at the majority of points, except in the areas of the the mental, supraorbital, upper molar, midmandible, lower cheek, lateral nostril, infraorbital, lateral orbit and supraglenoid points, which were similar for men and women. Women of all ages had significantly more tissue than men at the zygomatic attachment, and women between 35–45 years had significantly more tissue at the subcanine point (see Table 5.8).

In conclusion, the cadaver and live studies of White European, White American, Black American, Mixed Race South African, and nine ethnic groups from the Soviet Union, Indian and Japanese populations are all in agreement. The results suggest that men have thicker tissues over most of the face than women, especially at the brow, mouth and jaw. Women have thicker tissues at the cheeks.

Differences in facial tissues related to age

Differences between juvenile and adult facial tissues will be discussed in Chapter 8. Adult age-related tissue changes are apparent from some of these studies. His (1895) found that, with increased age, White European men showed increased tissues at all areas except the nasion, midphiltrum and gonial points (see Table 5.1). Czekanowski (1907) suggested that tissue increased with age in men, until 50 years, and then decreased (see Table 5.2). Suzuki (1948) agreed with His that tissues in men and women increase with age, especially beneath the chin. However, Suzuki did state that the tissue over the masseter muscle appeared to decrease between the ages of 60 and 80 years (see Table 5.3). Helmer (1984) showed that, as age increased in women, the tissue increased at the forehead and glabella, chin, cheek and jaw, and the back of the head, but decreased at the upper lip and temporal regions (see Table 5.7). He also showed that, as age increased in men, the tissue increased at the brow, nose, chin, canine fossa and jaw regions, and decreased at the nasion, mouth and lower cheek. The White American studies (Manhein *et al.*, 2000) showed that tissue increased with age at the glabella, lateral nostril, mental and supraorbital points, and decreased with age at the midphiltrum, supracanine, gonion and supraglenoid points (see Table 5.8). The same study showed that, in Black Americans, facial tissue did not increase with age at any areas, but decreased with age at the rhinion, midphiltrum, supracanine, gonion and supraglenoid points.

In conclusion, it appears that the age-related changes in tissues are very variable. However, these results do suggest that tissues at the mouth and lower cheek tend to decrease with age, and tissues at the chin and brow may increase with age.

Differences in facial tissues related to ethnic origin

A comparison of the studies by Birkner (1903–1907), Fischer (1903) and Von Eggeling (1909), and the combined White European results (see Table 5.3) showed significant tissue differences between different racial origin groups (see Fig. 5.3). Chinese faces had thicker tissue than Europeans and Papuans at the brow, upper lip, chin and zygomatic attachment. Papuan faces had thinner tissue than Europeans and Chinese at the nasion and zygomatic attachment, and thicker tissue at the midmandible and gonial points. Hereron (Black African) faces had significantly more tissue than the other groups at the upper lip and the supraglenoid point. The tissues at the masseter region were thinner in the White European group than in all the other groups. Birkner (1907) concluded that the soft parts of the face vary more than the skulls between racial groups. When the Japanese population was compared with these groups, Suzuki (1948) found that Japanese faces had less of a difference

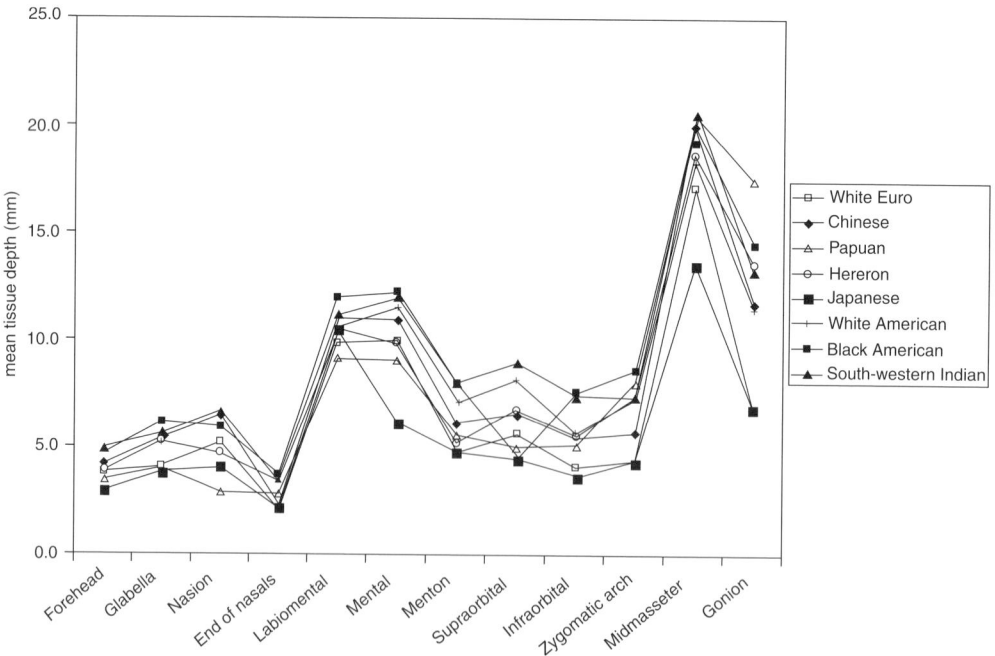

Fig. 5.3 Comparison of facial tissues between different ethnic groups. Modified from Table 5.3.

between the nasion and midnasal tissues (0.2 mm) than European (2.24 mm), Chinese (1.17 mm) and Hereron (1 mm) faces, but similar results to Papuans (see Tables 5.2 and 5.3). He also found that Japanese faces had thinner tissues at the chin, around the eyes and at the masseter areas than all the other groups, and thinner tissue at the zygomatic attachment than European, Chinese and Hereron faces. Japanese faces also showed thicker tissue at the supraglenoid point than all the other groups.

Rhine and Campbell (1980) found that Black subjects showed thicker tissue depths than European faces at all facial areas apart from the frontal eminence (see Table 5.3). The greatest differences appeared at the lips, beneath and lateral to the eyes, and at the mandible. This may have been due to the use of fresh, undistorted cadavers by Rhine and Campbell, or it may be due to an increase in stature and weight between the early part of the twentieth century and the 1980s. It may also be due to racial group differences between White Europeans and Black Americans. Rhine and his colleagues then compared the Black American results with White American results (Rhine et al., 1982). They found that the White American measurements fell between the White European and Black American measurements at the majority of points (see Table 5.3). The Black Americans had thicker tissues than White Americans at all points, except the forehead, nasion, rhinion, and lower first molar, where the tissues were similar, and the supraorbital point, where the tissues were thinner. Japanese faces showed smaller measurements at all points compared to Black American faces, and smaller measurements than both White Americans and Europeans at the majority of points. The exceptions were at the chin–lip fold, where the Japanese tissue was thicker than Whites, and at the zygomatic arch, where the Japanese tissues were similar to Whites. Rhine (1983) found that South-western Indians showed thicker tissue at all areas than the Japanese. South-western Indians showed tissue depths somewhere between Blacks and Whites, except below the lower molar region, where the tissue was thicker in South-western Indian men than both Black and White men (see Table 5.3). Phillips and Smuts (1996) suggested that

Black faces have thicker tissue than Mixed Race faces at the upper and lower parts of the face, notably at the lower lip, frontal eminence and cheeks (see Table 5.6). They also found that Mixed Race faces had thicker tissue at the philtrum, upper lip and gonial areas than White faces, but thinner tissue at the nasion, mental, supraorbital, lateral orbit and cheek points. Phillips and Smuts suggested that Mixed Race faces are not midway between Black and White faces, but should be considered a unique spectrum.

Lebedinskaya and her colleagues (1993) found wide intergroup variation and minor differences between the nine different Russian ethnic groups that she studied (see Table 5.9). There was some overlap between Mongoloid (Koreans, Abkhazians and Buryats) and Europoid (Russians, Lithuanians, Armenians and Uzbeks) groups, and one group of mixed origin – Kazakhs – approached the Mongoloids, whilst another – Bashkirs – approached the Europoids. A tendency for thicker soft tissues in the nasal region was noted in both sexes of Europoids as compared to Mongoloids. When the White European (Helmer, 1984), White American (Manhein et al., 2000) and White Australian (Simpson & Henneberg, 2002) ultrasonic measurements are compared, they appear very similar (see Tables 5.4, 5.7 and 5.8). The exceptions are that the male European faces had more tissue at the mouth and less at the gonion than the other groups, and the male Australians had more tissue at the glabella and nasal bones than the other groups. The female Europeans had more tissue at the mouth and beneath the chin, and less at the gonion than the other groups, the female Australians had more tissue at the nasal bones than the other groups, and the female Americans had less tissue at the glabella and alare than the other groups. Indian faces (Sahni, 2002) showed thinner tissues at the majority of points than White European, Black American, White American and Mixed Race South African faces (see Tables 5.3, 5.5 and 5.6). The Japanese faces showed thinner tissues at all points than the Indian faces. The exceptions are that Indian, White European and White American men had more tissue at the supraorbital point than Black American and Mixed Race men. Indian men had thicker tissues at the nasion, lower lip and upper molar regions than the Mixed Race men, and Indian women had thicker tissues at the lower lip and upper molar regions than the Mixed Race women. Japanese faces had similar tissue to Indian faces at the rhinion and chin–lip fold.

In conclusion, there appear to be great differences in facial tissue depths between different ethnic groups. This may be a reflection of racial origin group variation, ethnic group variation, stature and weight, or just a reflection of the enormous variation between individuals in a world population. The most consistent variations are apparently that African faces have thicker tissue in general than other groups, especially at the mouth, beneath and lateral to the eyes, and at the mandible. Japanese faces have thinner tissues than most other groups, except at the chin–lip fold and side of the head. Whites and Indians have thicker tissues at the supraorbital region than Africans and Mixed Race individuals. Mixed Race faces have thinner tissue than African faces at all points, and thinner tissue than White faces at most points, except the upper lip, which has thicker tissue for Mixed Race than White faces. South-western Indians have tissue depths somewhere between Whites and Blacks, except at the lower cheek, where they have more tissue.

The importance of tissue depth measurements to the accuracy of facial reconstruction

The importance of these tissue depth measurements during the facial reconstruction procedure has been debated. Gerasimov suggested that the facial reconstruction technique should rely only upon anatomy, and he stated that the skull alone provides the information necessary to recreate the face. Indeed, it can be argued that the tissue depth data only represent mean tissue measurements, and that these cannot be relevant to all skulls within the same sex, age and racial origin groups. In fact the relevance of average information in a field that relies upon the idiosyncrasies of individual faces must be questioned. In addition, the use of erroneous sets of data may lead to inaccurate facial reconstructions.

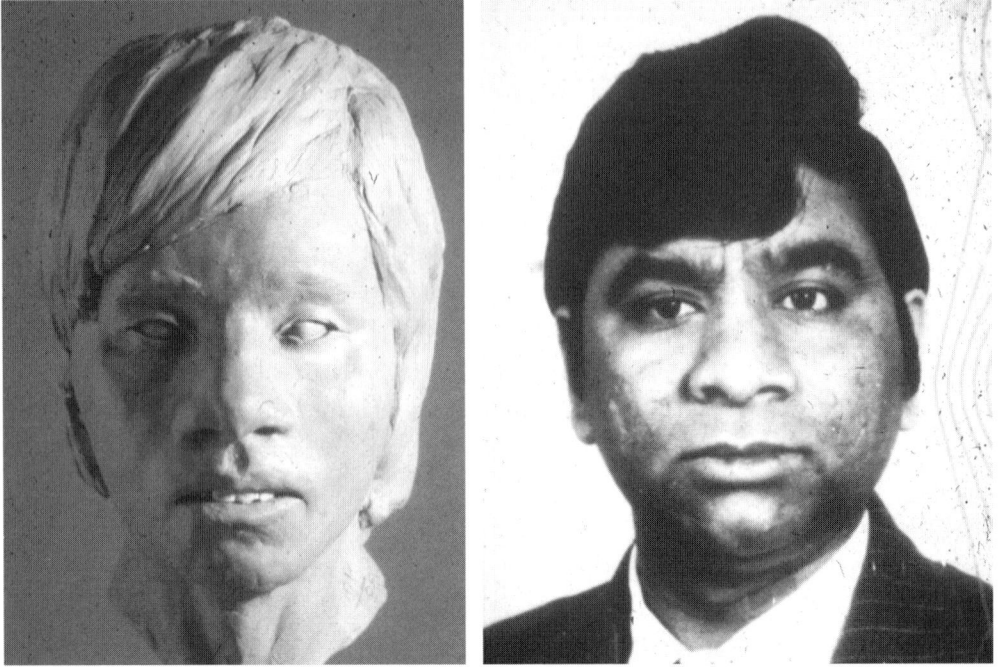

Fig. 5.4 Forensic facial reconstruction by Neave (Prag & Neave, 1997). The facial reconstruction (left) was produced as a young Chinese or Malaysian man in his twenties. The identified individual (right) was an Indian man in his forties.

Not infrequently, forensic science may suggest an age, sex or racial origin group that is later proved to be incorrect (see Fig. 5.4). Since sex determination in juveniles can be little better than guesswork, racial origin determination in adults can only be 80 per cent accurate (and less so in children), and age determination, without microscopic dental analysis, may be up to 20 years out, mistakes in all three basic factors are entirely plausible. Under these circumstances the facial reconstruction may be compromised by the use of incorrect tissue depth data. A previous study (Auslebrook & Van Rensburg, 1982) produced three reconstructions of the same skull, of ambiguous racial origin, using Caucasoid, Negroid and composite tissue depth data. The composite tissue data were determined by selectively combining the other two sets of data, as suggested by the skull morphology. Volunteers favoured the composite face (70 per cent) over the Caucasoid (6 per cent) and Negroid

(24 per cent) faces, when asked to choose a resemblance to the photograph of the identified Mixed Race man. This result suggested that inappropriate facial tissue data may compromise the accuracy of the facial reconstruction. In addition, in archaeological cases, the tissue depth data will be from a different period of time to the remains. Clearly we do not possess Ancient Egyptian or medieval British tissue data, and this problem may also have an effect upon the accuracy of the facial reconstruction technique.

The problem has been studied at the University of Manchester (Wilkinson *et al.*, 2002) where the same skull was reconstructed using six different sets of tissue depth data from six different ethnic origin groups. Six plaster copies of the same skull were studied. The skull was originally a stereolithographic copy produced from computed tomography data of a known, live male volunteer (Man X) (see Fig. 5.5). Man X was White European and over the age of 60 years. Six 3-D facial reconstructions of the

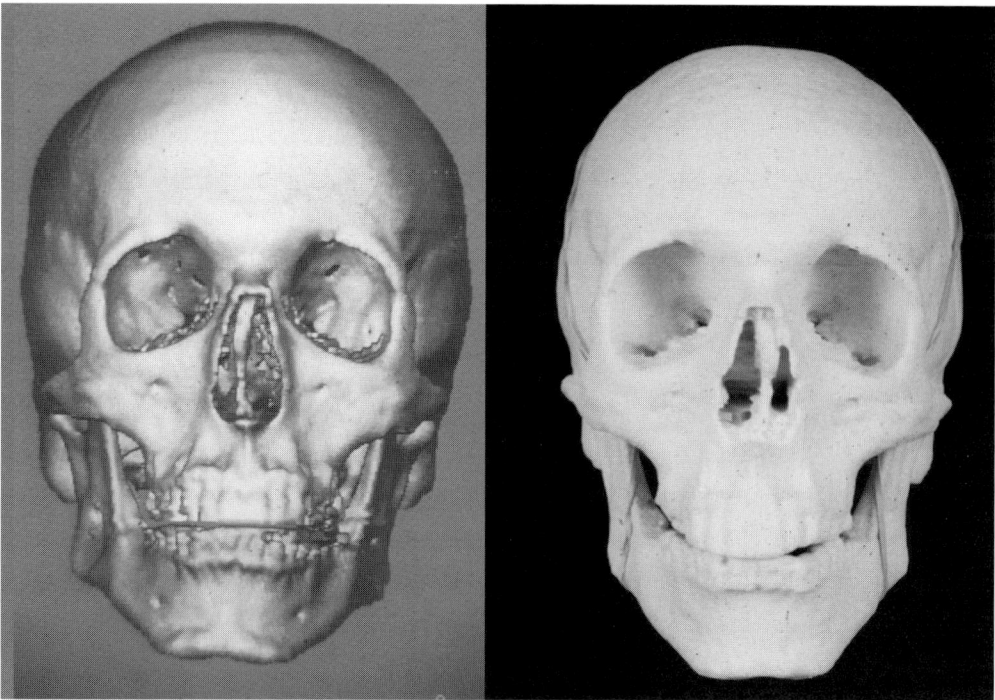

Fig. 5.5 CT scan (left) and copy of the skull (right) of Man X.

skull were produced, by three researchers, following the same methodology. Prior to reconstruction, the skull was assessed with respect to facial morphology, and the three researchers followed the same facial feature guidelines for each reconstruction. The skull copies were mounted in the Frankfurt Plane. Holes were drilled into the skull at 90° to the surface, at the appropriate number of anatomical points determined by the chosen set of tissue depth data, and a different set of data were used for each skull. These included: White European (Helmer *et al.*, 1989), Black American (Rhine & Campbell, 1980), Korean (Lebedinskaya *et al.*, 1993), Japanese (Suzuki, 1948), Mixed Race (Phillips & Smuts, 1996) and South-western Indian (Rhine, 1983) (see Tables 5.3, 5.5, 5.7 and 5.9). Wooden pegs were cut to the lengths governed by the tissue depth data and inserted into the holes in the skulls. Using this method sets of guides for tissue depth across the face were attached to the skull surfaces. The muscles of the face were modelled in clay onto the skulls one by one following the Manchester methodology. The details of the facial features (nasal shape, lip form and eyebrow pattern etc.) were modelled with respect to the prior assessment of the skull. The finished reconstructions were photographed in frontal and profile views from eye level (see Fig. 5.6). A total of 247 volunteers were then selected from a student population. The volunteers were shown images of Man X in frontal and profile view and were asked to study the images for five minutes (see Fig. 5.7). The images were then replaced with images of all six reconstructions in frontal and profile views and the volunteers were asked to rate each reconstruction as a resemblance of Man X. The ratings were made on a scale of 0 to 10, with 0 = no resemblance, 5 = moderate resemblance, and 10 = great resemblance. The White European and the Mixed Race heads were rated as the closest resemblance to Man X by 36.8 per cent of the volunteers, and they both recorded mean resemblance rates of six (moderately good). The South-western Indian and the Japanese heads recorded mean resemblance ratings of 4.8 (moderate) and were rated as the closest resemblance to Man X by 23.9 per cent and 16.2 per cent of the volunteers respectively. The Black American and Korean heads

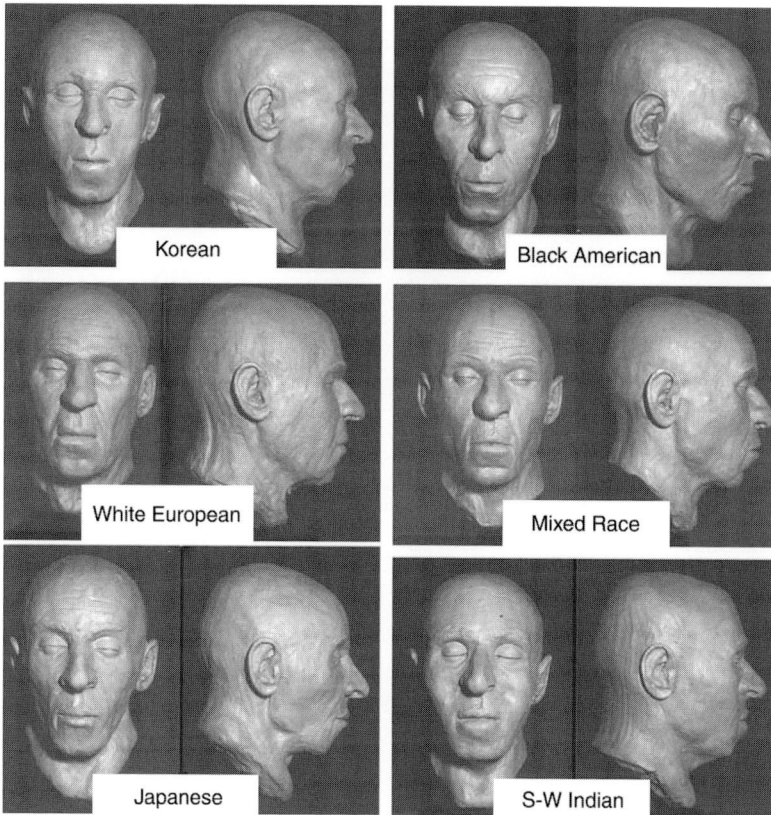

Fig. 5.6 Facial reconstructions produced using different ethnic group tissue data.

recorded mean resemblance ratings of 4.1 (approximate) and were rated as the closest resemblance to Man X by 12.6 per cent and 8.9 per cent of the volunteers respectively (see Table 5.10). The Mixed Race and White European results were not significantly different from each other, and this also applied to the South-western Indian and Japanese results, and the Black American and Korean results. All the heads recorded mean ratings between four and six (approximate to a moderately good resemblance to Man X). Sixty per cent of the volunteers stated that the reconstructions were similar to each other and 6 per cent stated that there were no similarities between the reconstructions. Thirty four per cent stated that some of the reconstructions were similar to each other, and the most

Table 5.10 Resemblance assessment results for the six facial reconstructions.

	Facial reconstructions						
Ethnic group	Korean	White Euro	Japanese	Black American	Mixed Race	S-W Indian	All
Question 1							
% best likeness	8.9	36.8	16.2	12.6	36.8	23.9	
Mean resemblance	4.1	5.9	4.8	4.1	6	4.9	5
Standard deviation	2.14	2.14	2.29	2.23	2.11	2.63	2.23
Wilcoxon tests							
Korean		<0.001	0.001	0.94	<0.001	<0.001	
White European			<0.001	<0.001	0.787	<0.001	
Japanese				0.001	<0.001	0.226	
Black American					<0.001	<0.001	
Mixed Race						<0.001	
Question 2							
% all similar							59.5
% some similar							34.4
% not similar							6.1

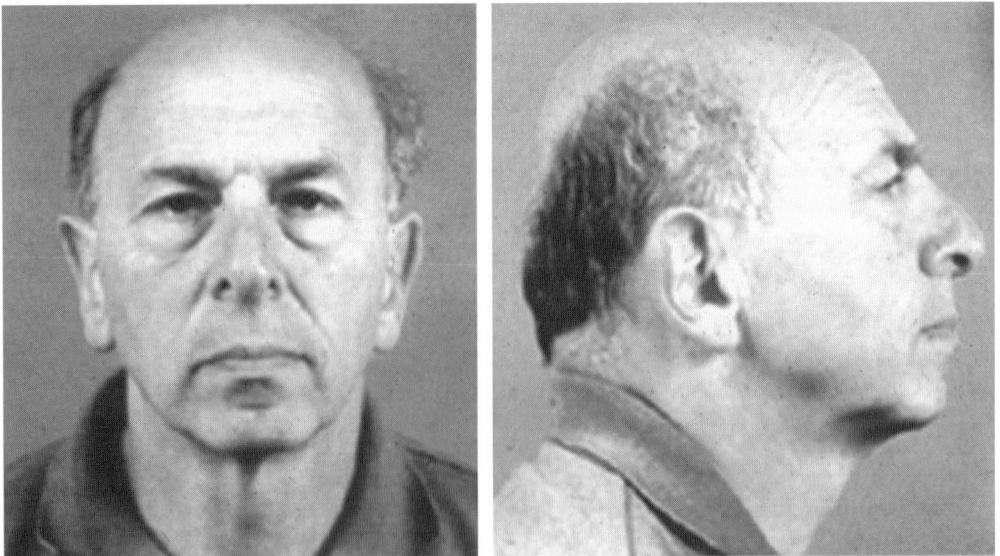

Fig. 5.7 Man X.

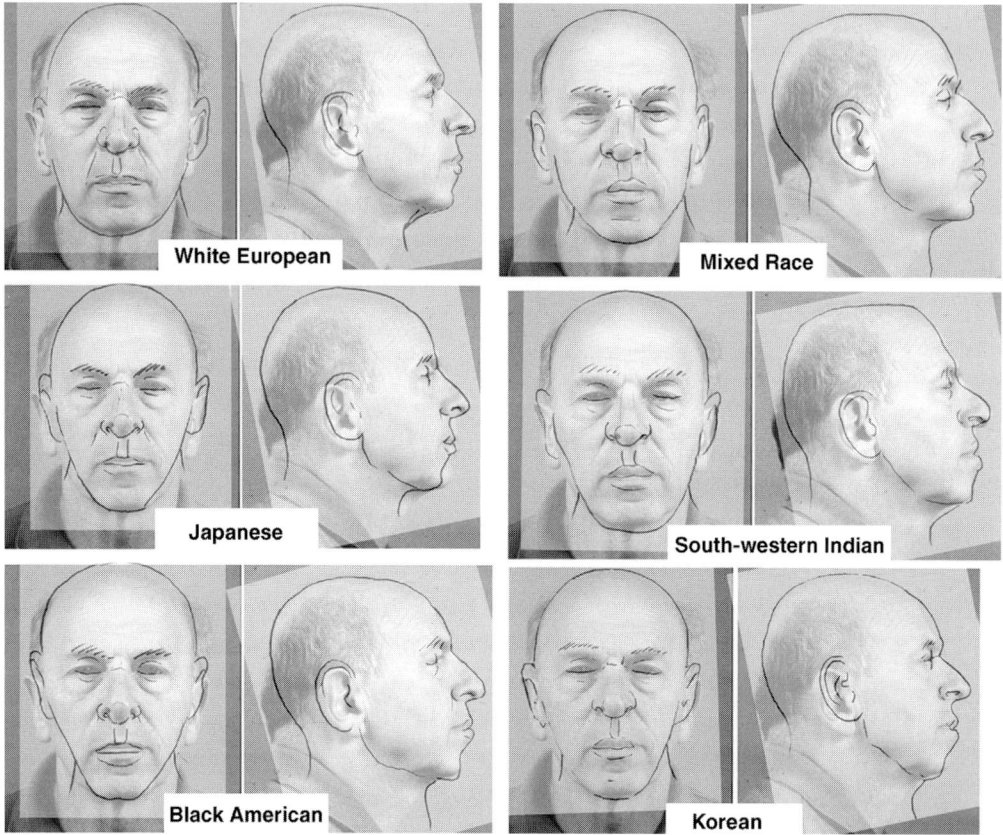

Fig. 5.8 Photographic superimposition of Man X with outline tracings of all the facial reconstructions.

frequently chosen were the Mixed Race and White European heads, then the Korean and Japanese heads, and finally the Korean and Black American heads. Photographic superimposition assessments showed inaccuracies in all the reconstructions, but the White European and Mixed Race heads showed the least inaccuracies (see Fig. 5.8). The facial tissue differences appear to have affected the resemblance of the other four reconstructions of Man X, but even the heads with the greatest tissue depth differences compared to the White European head (i.e. the Black American and Japanese heads) still recorded approximate resemblances to Man X. All the reconstructions showed consistent inaccuracies at the nasal tip shape, and the ear shape, size and projection. These inaccuracies can be considered as problems with the reconstruction technique as the skeletal detail cannot indicate the nasal tip shape or ear pattern. The chin of Man X also showed a more definite cleft and squarer shape than on all the reconstructions, and all the reconstructions showed a deeper nasion than Man X. This must be a reflection of individual variation, which cannot be accurately determined using the average tissue depth data. These results suggested that correct ethnic origin tissue depth data are important in facial recognition, but that a reasonable resemblance can be produced even with an incorrect set of data. The tissue depth data clearly affect the facial appearance, but the skull itself was shown to be the most important factor when producing a facial reconstruction. The correct set of tissue depth data will produce the most accurate likeness, but erroneous tissue depth data will not affect the face to the extent that it cannot be recognised by a relative or close friend. However, these tissue guides are very important for the facial reconstruction of children, since the age of a child can have a very significant effect upon the soft tissue distribution on the face. The use of adult data for an eight-year-old child, for example, would always be inappropriate (see Chapter 8).

Strict adherence to the exact tissue depth measurements during the facial reconstruction procedure should be avoided. Experience suggests that the tissue data may often suggest depths that appear inappropriate for the skull, even when the data are from the correct sex, age and ethnic origin groups. In these cases the direction of the skull and the anatomy should be followed, with tissue depth measurements being used only as guides.

6 The Manchester method of facial reconstruction

The production of a facial reconstruction requires intimate knowledge of facial anatomy, and considerable research should be carried out prior to any reconstruction attempt, preferably using human dissections. Excellent facial anatomy references can be found in Gray's Anatomy (1980); Sobotta (1983); Kahle *et al.* (1992). Figs. 1.13 and 6.1 illustrate the facial musculature. Forensic facial reconstruction cases tend to involve human remains that cannot be identified by the usual police investigative channels. The police may have already followed all crime scene clues such as physical, circumstantial or biological evidence. The identification of an individual is the field of the forensic anthropologist and the forensic odontologist, and many routes of analysis may be pursued. Analysis of the skeleton determines the sex, age, racial origin, stature and pathological condition. A written description of the individual may not elicit a response from the general public or the authorities, and a facial reconstruction may be suggested as a last option. Usually in these cases the original material is skeletal, but soft tissue may be present if the body has been burnt, partially decomposed or submerged in water. If any soft tissue is present this may provide important information regarding the individual's facial features and, ideally, the remains should be assessed and photographed prior to receipt of the skull for reconstruction.

Skull assessment

Whether forensic, archaeological or clinical, the starting point for a facial reconstruction is always the same: the skull. The skull can be

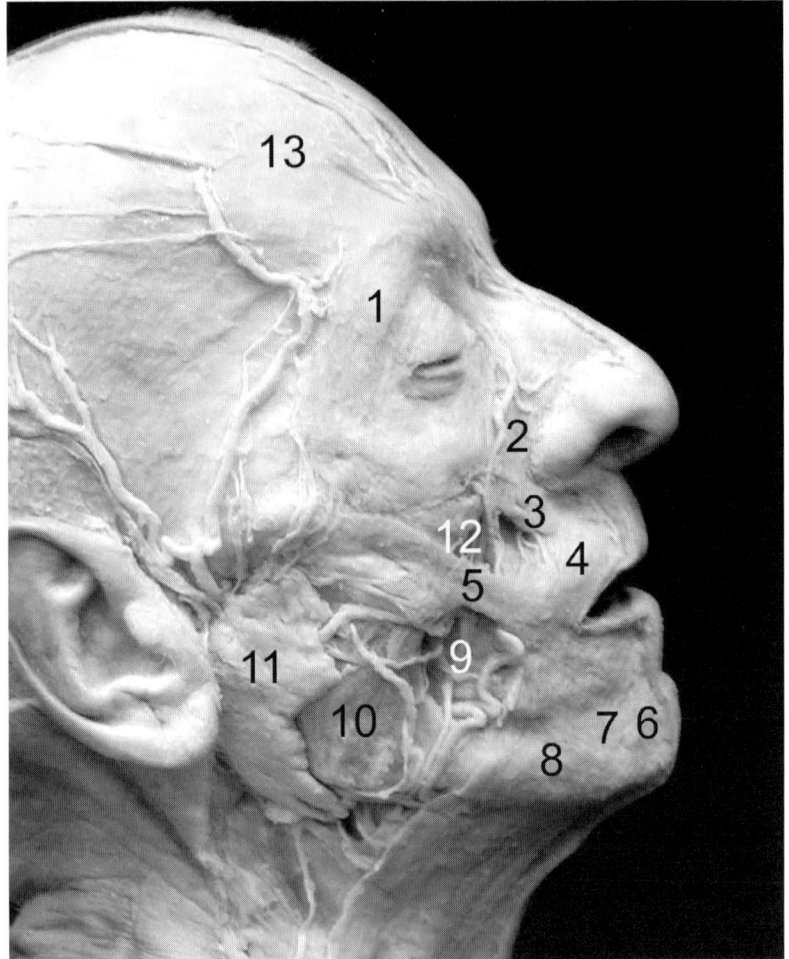

Fig. 6.1 Facial musculature.

1	Orbicularis oculi	8	Depressor anguli oris
2	Levator labii superioris alaeque nasi	9	Buccinator
3	Levator labii superioris	10	Masseter
4	Orbicularis oris	11	Parotid gland
5	Zygomaticus major	12	Levator anguli oris
6	Mentalis	13	Frontalis
7	Depressor labii inferioris		

quite fragile in parts and should be handled above a soft surface, such as a foam sheet or padded layer, and stabilised with sandbags, beanbags or foam to prevent rolling. Care should be taken when handling the skull and both hands should always be used. The

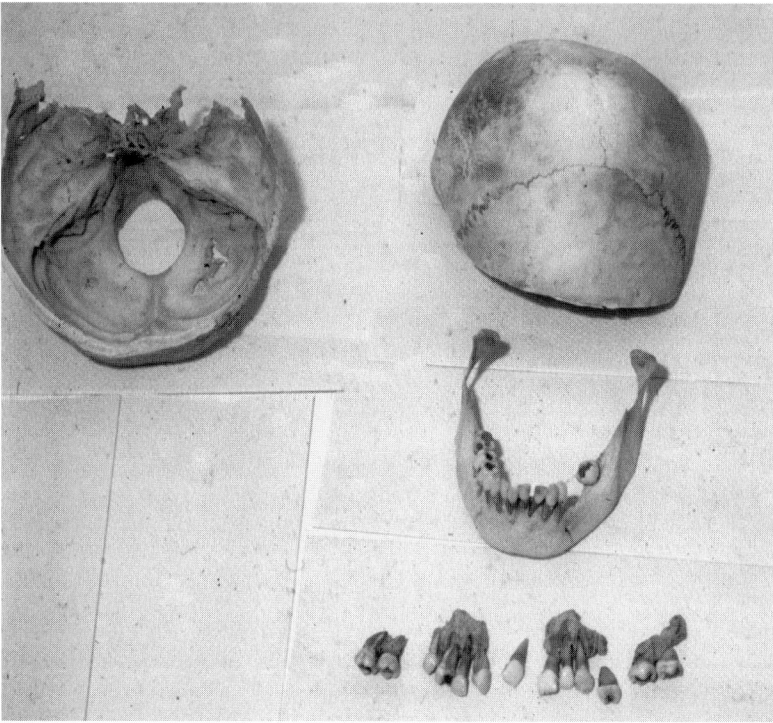

Fig. 6.2 Fragmented skull.

orbital bones, nasal bones, sinus bones and zygomatic arches may be damaged easily and should not be used to grip the skull. Frequently the skull is presented incomplete and fragmented (see Fig. 6.2), with some missing, damaged or broken bone fragments. The skull may be damaged deliberately to attempt to pervert the course of identification, accidentally during decomposition or when the body is found, or the skull may have been damaged during its burial period. In these cases the skull must be re-assembled before the facial reconstruction procedure can continue. Sticky wax or vinyl acetate can be used to glue the skull fragments together (see Fig. 6.3), and these materials are optimal for re-assembly since reshuffling of the fragments can be achieved by loosening the adherence (heating the wax or pulling the acetate) without causing damage to the skull fragments. A great deal of reshuffling may be necessary during skull re-assembly. It is a mistake to begin

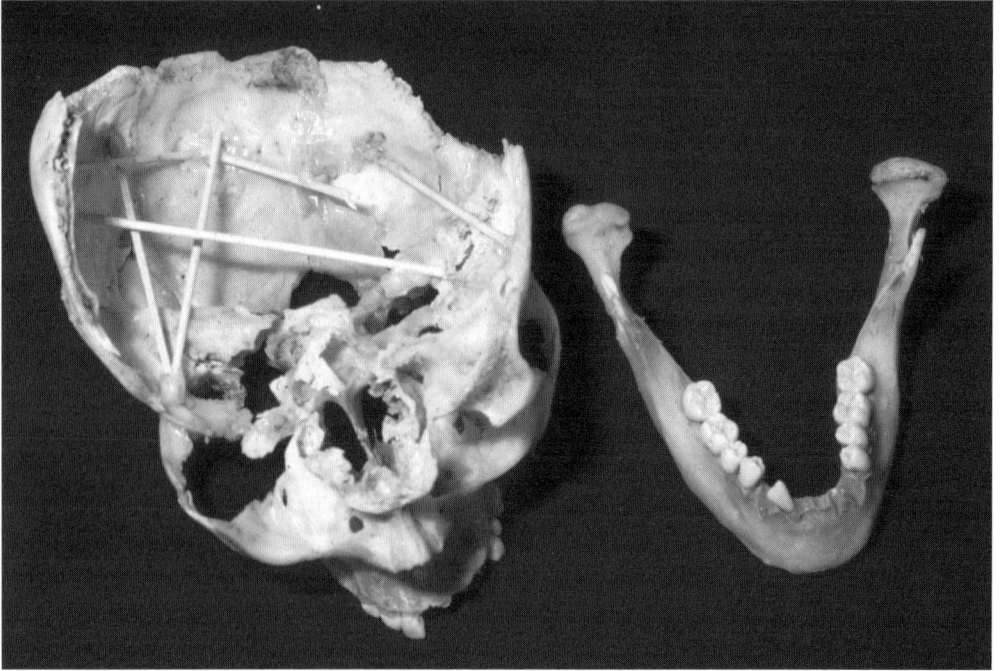

Fig. 6.3 Re-assembly of a fragmented skull.

re-assembly at one point on the skull and continue adding fragments without reassessing the re-assembly, since any errors in
alignment will produce a distorted result. A small error in alignment will be greatly magnified throughout the whole skull. It may
be necessary to seek the advice of a pathologist or paleopathologist to establish whether any distortion is natural, postmortem
or due to a pathological condition. Intact skulls may hold a degree of tension within the vault of the skull, and when the skull
fragments it may become 'unsprung'. This release of inbuilt tension may mean that it is impossible to re-assemble the skull to
its original form, and a degree of compensation may be necessary. Since the facial bones are the most influential in the facial
reconstruction procedure, any compensatory overlap of fragments
should be considered at the back of the vault, where the errors
in the shape should not affect the accuracy of the reconstruction.

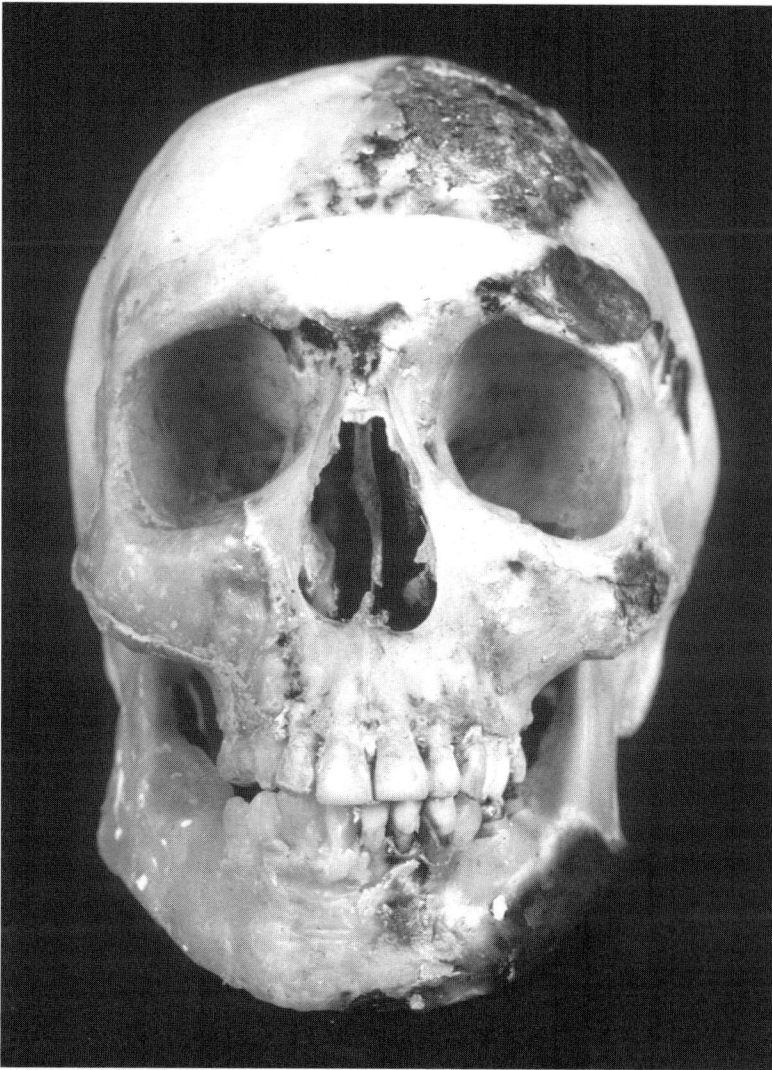

Fig. 6.4 Skull with mirror-image remodelling of missing areas (left).

Any missing areas are modelled in wax onto the skull (see Fig. 6.4). Unilateral missing areas can be modelled from the available area on the other side of the skull, so that a mirror image is created. It must be noted that very few skulls are symmetrical and this modelling technique creates some errors, since the resulting skull will have a false symmetry. Since most skulls are asymmetrical, most faces are also asymmetrical. The degree of asymmetry in the face can be illustrated with the technique (Gerasimov, 1975) detailed in Chapter 4 (see Figs. 4.6 and 4.7). However, the composite images do suggest that mirror-image modelling of missing areas should not greatly affect the accuracy of the reconstruction, since most people would still be recognised from one of the composite images. Where the missing areas of the skull are from unilateral features, or where both sides of the skull have absent areas, the areas are estimated using the surrounding bones as guides (see Fig. 6.5). Research (Colledge, 1996) at the University of Manchester suggested that areas of the skull can be estimated with relative accuracy. Colledge took five skulls and attempted to remodel a different missing area on each skull in a blind study. The missing areas included the mandible, the frontal bone, the zygomatic bones, the maxilla and the occipital bone (see Fig. 6.6). The remodelled skull was then compared metrically with the original specimen. Colledge found that the modelled areas were not significantly different from the original parts, except at the mandible. The mandible was remodelled with substantial errors, especially at the jawline and chin height. This would have a significant effect upon the accuracy of any resultant facial reconstruction.

When the cranial fragments have been re-assembled and any missing areas remodelled, the skull detail can be examined and a descriptive record of facial morphology produced. Included in the following records are those standards routinely employed for the Manchester method, taken from some of the research results overviewed in Chapter 4. Any asymmetries should be noted, and the following points recorded:

1 The overall face shape

The facial form can be determined from the cranial shape and gonial angle. The transverse arc of the cranium will indicate round, square, oval or triangular face shapes (see Fig. 4.8). If the gonial

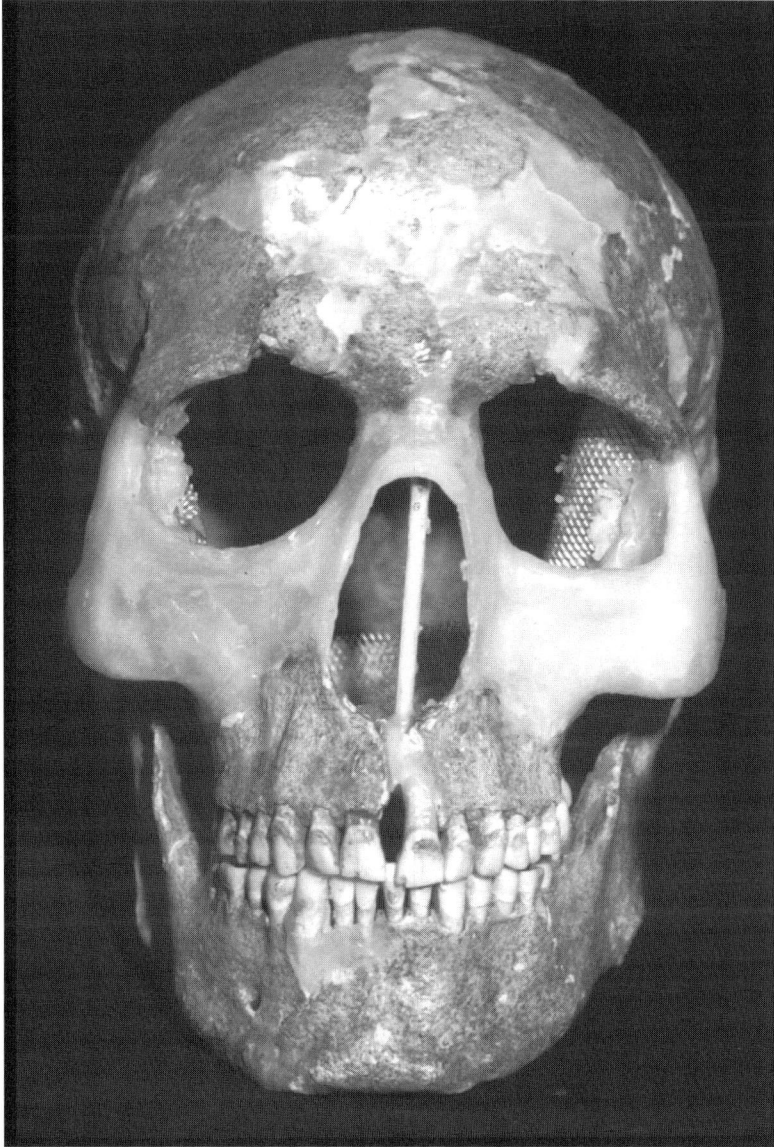

Fig. 6.5 Remodelled skull.

angle is more than 125° with a high coronoid process on the mandible, the face will be oval or triangular, and if the gonial angle is less than 125° with a wide and low coronoid process, the face will be round or square.

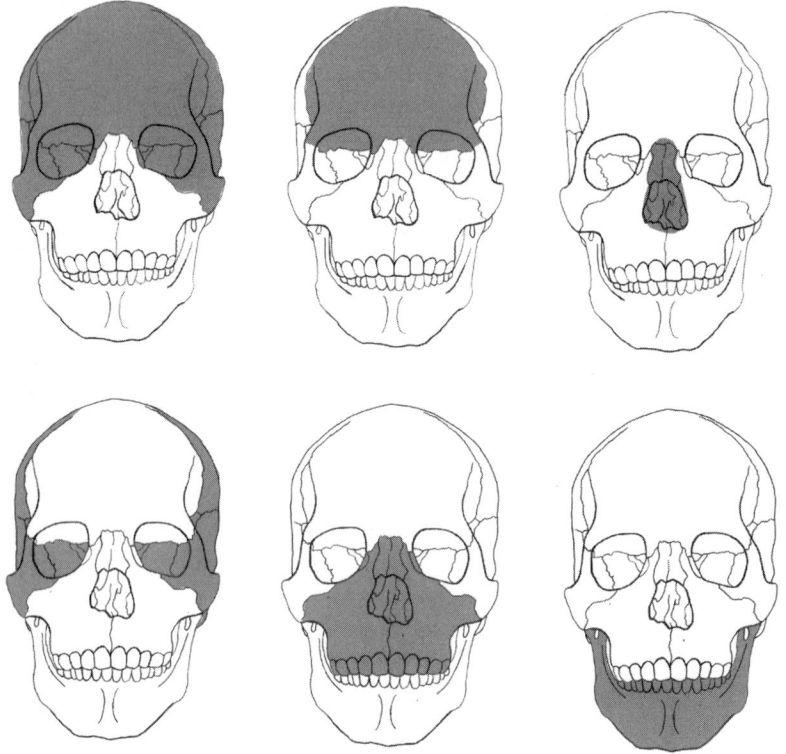

Fig. 6.6 Missing areas of skull as studied by Colledge (1996). Six skulls were studied. Missing areas were blindly remodelled (areas in grey) and then compared to the original specimen. All areas were successfully remodelled except for the mandible, where quite large errors occurred.

2 Hairline

The hairline may be seen microscopically where the smooth surface of the bone changes to become rough with small tubercles.

3 Forehead creases

The muscle attachments at the brow suggest the presence of forehead creases. Strong attachments of the procerus muscle suggest a nasoglabellar crease, and strong attachments of the corrugator supercilii muscles suggest vertical glabellar lines. Strong attachments of the occipitofrontalis muscle suggest horizontal forehead creases, but the pattern of these creases cannot be determined.

4 The brow shape

A deep nasion and deep-set eyes suggest a nasoglabellar crease. The strength of the supraorbital ridges will determine the degree of brow ridges.

5 The eye fissure

A tangent between the malar tubercle and lacrimal crest will determine the slope of the eye fissure (see Fig. 4.12). The eye slit is 60–80 per cent of the width of the orbit, with the eyeball located in the centre of the orbit.

6 The eyeball

The depth of the orbits suggests the eyeball projection. To some extent this is a matter of experience. Deep-set eyes are suggested when the supraorbital rim is thickened and protrudes more than the infraorbital rim. Wide, protruding eyes are indicated by a weak orbital profile and a smooth, thickened lateral supraorbital rim. The eyeball should be positioned so that the flat plane of the iris is touching a tangent taken from the mid-supraorbital point to the mid-infraorbital point (see Fig. 6.7).

7 The eyelid pattern

A low nasal root plus a strong anterior lacrimal crest suggests the presence of the medial epicanthic (Mongoloid) fold. A low orbit, with an overhang of the brow ridge, suggests an intermediate plus a lateral epicanthic fold. The fixed upper eyelid fold generally follows the direction of the supraorbital margin (see Fig. 4.13). If there is an overhang in the middle of the supraorbital rim, the fold of the eyelid is located in the middle of it. If the lateral rim is thickened, the eyelid fold is more defined laterally. A high orbit, a low or medium-height nasal bridge and a long lacrimal fossa indicate that the fold is more defined medially (see Fig. 4.13).

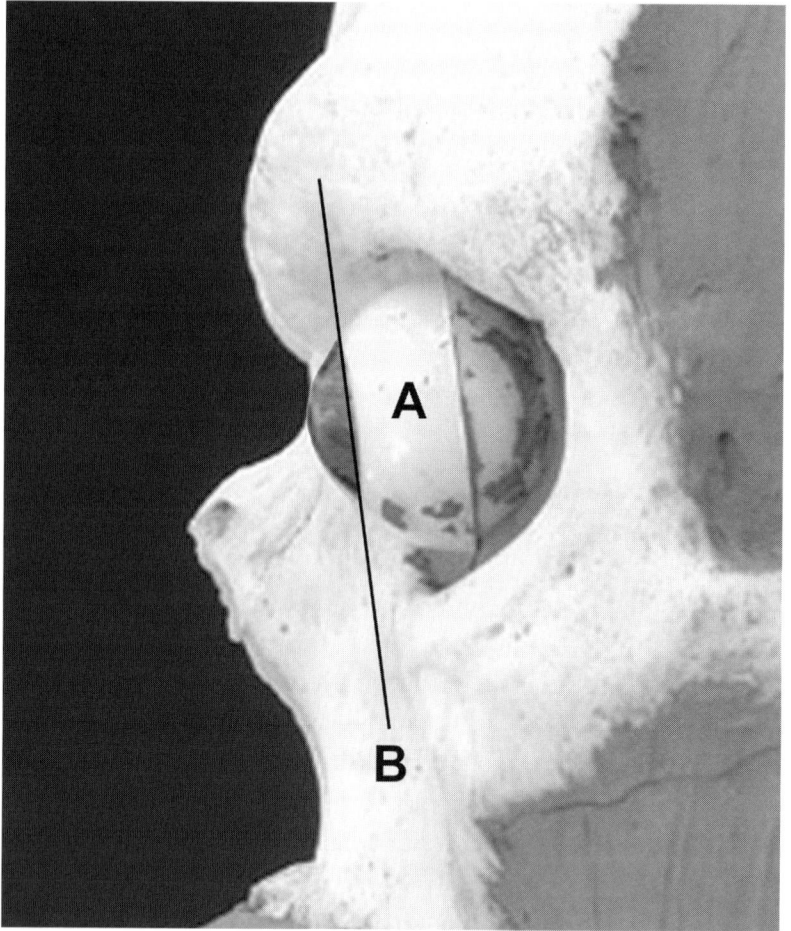

Fig. 6.7 Normal eyeball protrusion (Wilkinson & Mautner, 2003). A = eyeball, B = tangent from mid-supraorbital to mid-infraorbital points.

8 The eyebrow pattern

The eyebrows generally follow the line of the brow ridge and they will be approximately 3–5 mm above the supraorbital margin (see Fig. 4.14). A strongly developed supraorbital margin and brow ridge indicate that the eyebrows are positioned 1–2 mm lower than the supraorbital rim. A low nasal root and weak brow ridges suggest that the medial third of the eyebrow is located beneath the supraorbital rim, whilst the lateral two-thirds of the eyebrow slopes up to the supraorbital margin and then traces its contour. If there is thickening of the lateral supraorbital rim and a strong brow ridge, the eyebrow follows a triangular line. A high nasal root and smooth brow suggest arched eyebrows.

9 The nasal projection

The projection of the nose is determined by two straight lines, one following the last portion of the nasal bones, and the other as a continuation of the main direction of the point of the bony spine (see Fig. 4.10). The point of intersection of these two lines will give the position of the tip of the nose. This does not suggest the profile of the nose or the slope of the dorsal ridge, merely the most anterior point of the nose. Care must be taken when assessing the nasal bones as there is a tendency for the bones to change direction at the end of their contour, and this new direction can easily be missed, leading to a nasal projection that is too pronounced. Where the nasal spine is damaged or absent, the line of projection can be determined from the slope of the bony palate, which is parallel to the slope of the nasal spine.

10 The nasal width

The width of the nasal aperture, at its widest point, is three-fifths of the overall width of the nose (at its widest point).

11 The nasal spine

The slope of the nasal spine reflects the slope of the nasal base (an upward-sloping spine leads to an upturned columella), and the form of the nasal spine determines the nasal tip. A spatulate nasal spine leads to a wide or bulbous tip, and a bifid spine leads to a cleft nose (see Fig. 4.9).

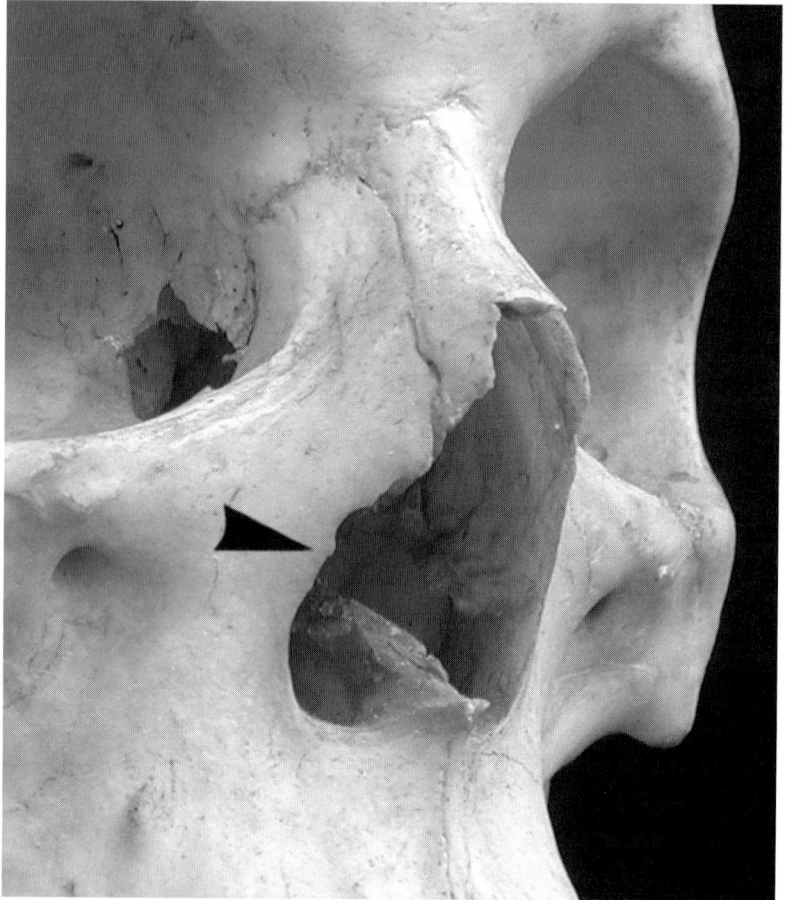

Fig. 6.8 The change in direction of the piriform contour.

12 The alar shape and position

The lateral spread of the piriform aperture sets the alar width and
the alae bulge to one-fifth of the overall nasal width on each side
of the lateral piriform edge. Clearly, asymmetries must be taken
into account. The height of the wings of the alae is influenced
by the position of the crista conchalis in the nasal opening. The
lateral nasal bones often exhibit a sharp change in direction along
the piriform contour, and this point will indicate the height of the
alae (see Fig. 6.8).

13 The nasal profile

A straight, thin nose usually has a weak glabella, a small inter-orbital distance, a narrow and high nasal bridge, an elongated narrow nasal aperture with simple contours, a high roof-like dorsal part of the nasal bones, sharp lower angles of the aperture, a nasal spine which is never directed downwards and is usually directed forward, thin nasal bones and a sharp facial profile. A hooked or hawk nose is described as being characteristic of a face with a strong glabella, sharp and prominent nasal bones, a narrow base, thick bell-shaped nasal bones, a symmetrical aperture, a nasal spine which is strongly developed and directed forwards or downwards (when the spine is directed downwards the nose is bent and bill-like) and a narrow face. A fleshy, broad nose is described as being characteristic of a broad roundish aperture, broad rounded nasal bones with simple contours and a bell shape, forward or upward-directed nasal spine and guttering of the inferior border. The snub nose usually has a short aperture, dull and rounded lower border, short upward-directed nasal spine and a concave roof with a wide and rounded arch. Any nasal bone asymmetry will also be present in the external nose (see Fig. 4.9).

14 The nostril position

The shape of the inferior borders of the nasal aperture will suggest the position of the nostrils. When there is nasal guttering at the inferior border, the nostrils will be visible in the Frankfurt Plane, and are placed laterally. When the inferior border is sharp and complete, the nostrils will also be placed laterally and will be visible in the Frankfurt Plane. When the inferior border is sharp and is divided along its contour, the nostrils will be placed inferiorly and will not be visible in the Frankfurt Plane.

15 The lip thickness

The thickness of the lips is based on the prognacy of the teeth, the incisors and alveolar parts of the upper and lower jaws. Small straight teeth are characteristic of thin lips and orthognathism. Prominent big teeth are characteristic of thick lips and prognathism. The height of the enamel of the central incisors is related to the thickness at the middle of the pigmented part of the lip, but the relationship is dependent upon sex and ethnic origin. Lip thickness in White European faces can be calculated from teeth height by the following formulae:

upper lip thickness = 0.4 + 0.6 × (upper teeth height)

lower lip thickness = 5.5 + 0.4 × (lower teeth height)

total lip thickness = 3.3 + 0.7 × (total teeth height)

Lip thickness for Asians from the Indian subcontinent can be calculated from the teeth height by the following formulae:

upper lip thickness = 3.4 + 0.4 × (upper teeth height)

lower lip thickness = 6 + 0.5 × (lower teeth height)

total lip thickness = 7.2 + 0.6 × (total teeth height)

Prognathism in Negroids, Polynesians, Japanese and Malays would suggest thick lips, whereas prognathism in Caucasoids or Mongoloids is not accompanied by procheilia of the lips, and the mouth has a puckered appearance, with the upper lip leaving the teeth uncovered at rest. Mandibular prognathism suggests procheilia of the lower lip. The line of the lower edge of the upper lip is just above the middle of the incisor crowns. In Whites the vermilion shows a smooth junction with the upper lip, whereas in Blacks it is elevated to give the 'lip seam'.

16 The philtral width

The philtral width corresponds to the distance between the midpoints of the maxillary central incisors.

17 The mouth width

The corners of the mouth are positioned on radiating lines from the junction between the canine and the first premolar, or directly beneath the medial border of the iris.

18 The inclination of the mouth corners

The relative strength of markings for the levator and depressor anguli oris muscles determines the up or down placement of the corners of the mouth.

19 The occlusion of the teeth

If the teeth show an overbite, or maxillary prognathism, then the upper lip will project further than the lower lip. The lower lip will project further than the upper lip, if there is edge-to-edge occlusion or an underbite. An assessment of the teeth by a dental consultant or odontologist is recommended.

20 Lip shape

The lip shape may be determined by the dental pattern (see Fig. 4.15). Prominent maxillary canines suggest a wide, square upper lip shape, with a large proportion of the maxillary teeth showing when the mouth is at rest. The upper enamel line will suggest the upper lip shape: when the highest points are on the central incisors this suggests a cupid's bow shape to the upper lip, and when the enamel line is approximately horizontal this suggests a flat upper lip line with a philtral bow that is only slightly depressed. Prominent mandibular canines suggest a central depression to the lower lip and lateral lip fullness. The shape of the lower lip will follow the enamel line of the lower teeth.

21 The presence of a nasolabial fold

The nasolabial crease extends from the upper edge of the nostril towards the upper first molar, and its strength depends on the depth of the canine fossa, the degree of horizontal face profiling, the projection of the frontal surface of the cheekbones and the presence or absence of the teeth. The nasolabial fold is more pronounced when the canine fossa is deep (more than 5 mm), when the teeth are missing, and with advancing age. The markings for the origins of the levator labii superioris and zygomaticus muscles determine the curves and depth of the nasolabial fold and the possibility of a second lateral (buccal) crease.

22 The chin shape

The degree of elevation of the frontal part of the mandible and the width of its base define the width of the chin. If the height of the mandibular body decreases from the chin triangle to the side of the rami, then this forms a high chin. When the lower border of the jaw is softly rounded inwards, and has no crests and possesses no roughness, then the muscular tissue will gently cover the bone and the chin will have soft contours. When the lower border shows prominent crests then, correspondingly, there will be well-developed muscle and the chin will be heavy and massive in form. Strong attachments at the mentalis muscles and a divided mental eminence suggest a cleft chin.

23 The ear

If the supramastoid crest on the temporal bone is strongly developed and protrudes, then the ear will show upper protrusion. If the outer surface of the mastoid process is rough, then the ear will show lower protrusion. If all these attributes are in place, then the ear has total protrusion. If the mastoid processes are directed downward (with the skull in the Frankfurt Plane), the lobe is attached (adherent), and if the mastoid processes point forward, the lobe is free (non-adherent). Small mastoid processes, which are directed inside the skull, will suggest small ears that are close to the head, whilst massive prominent mastoids are characteristic of large spread-out ears. The angle between the axis of the ear and the angle of the nose has an orientation with the helix more anterior than the parallel.

24 The cheek shape

When the cheekbones are flat, the zygomatic muscle is located on a more frontal surface and when the cheekbones are strongly profiled the zygomatic muscles are located mainly on the side of the face.

25 The neck thickness

The size of the neck can be determined by the strength of the attachments of the sternocleidomastoid muscles and the trapezius muscle.

26 Trauma, disease or pathological condition

An assessment of the skull by a pathologist, paleopathologist or forensic anthropologist is recommended to determine any signs of trauma, disease or physiological conditions.

The casting procedure

The mould-making material is normal-set dental alginate ('Tiranti'). This material has a working time of two minutes and a setting time of three minutes, and is flexible whilst retaining great detail (see Fig. 6.9). The preferred reconstruction technique is to work directly onto a copy of the skull. This decreases the possibility of damage to the specimen, allows study of the skull throughout the entire reconstructive process, and leaves a record of the specimen after the reconstruction is dissembled. Some specimens are very porous and delicate and these have to be covered in a thin layer of aluminium foil, which is burnished onto the outer surface to protect the skull during the casting process. The mould-making procedure is not described in this publication, but it must be noted that this technique is complicated and time-consuming and requires thorough training and experience. Any attempt to make a cast of a skull without the necessary skills may result in damage to the specimen. Thorough descriptions of skull casting can be found in Prag and Neave (1997) and Clement and Ranson (1998). The skull copy is cast in Plaster of Paris and is measured against the original specimen

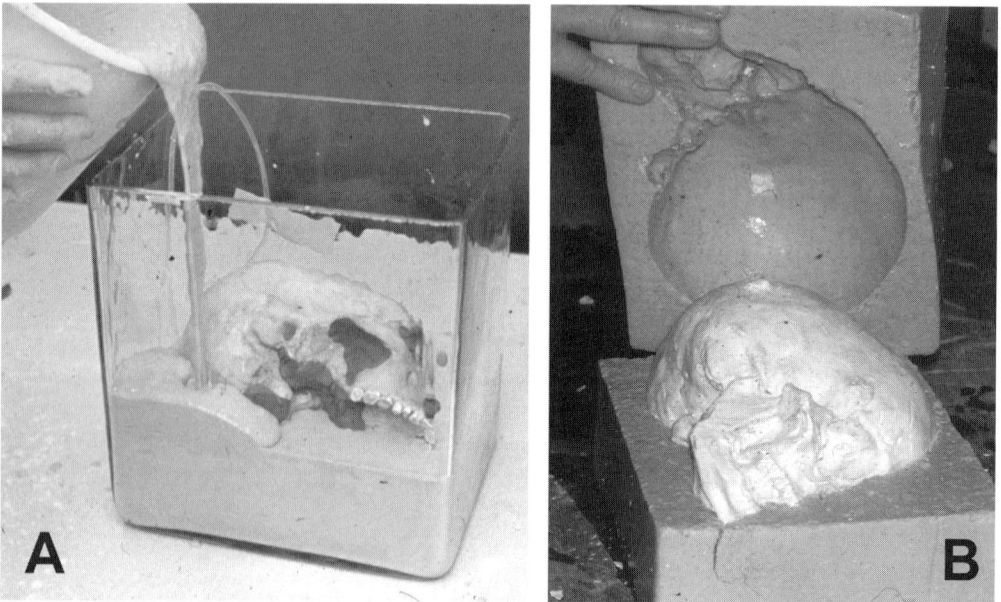

Fig. 6.9 Skull casting procedure. A = dental alginate poured around the original skull in two halves. B = original skull removed and plaster poured into the mould to produce a copy.

to check for any shrinkage or distortion during the casting procedure. There must be no demonstrable differences between the cast and the specimen for the cast to be accepted for use in the facial reconstruction procedure.

Facial reconstruction procedure

The skull copy is mounted onto a pole in the Frankfurt Plane. The Frankfurt Plane is reached when a horizontal line passes through the inferior border of the orbit and the auditory meatus on both sides of the skull (see Fig. 3.1). The mandible is attached to the cranium using dental sticky wax. The teeth are placed with the occlusion suggested by the wear of the teeth, and the condyles of the mandibles are attached at the temperomandibular joints with modelling wax. Before the wax is totally set, the teeth can be pried apart slightly to reflect the slack-jawed, relaxed position of the mouth that represented a living face. Where the skull is

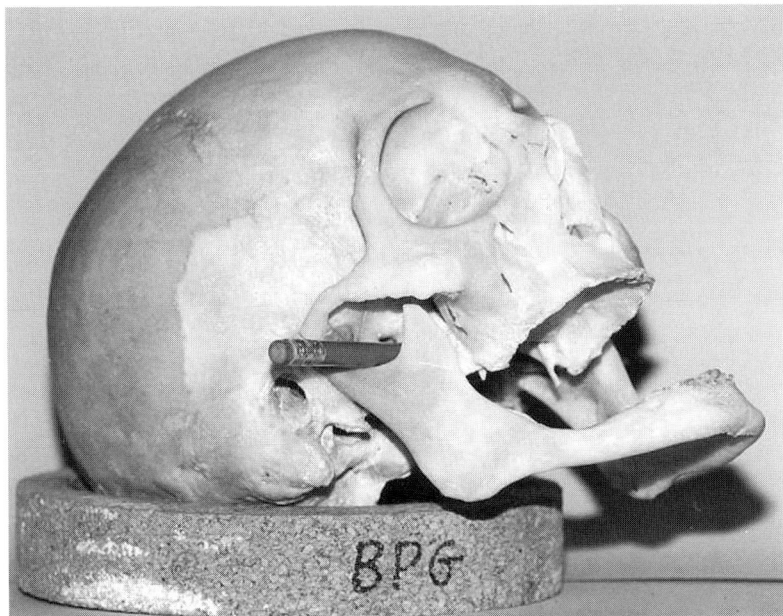

Fig. 6.10 The placement of the mandible in an edentulous skull. Courtesy of Karen Taylor (2001).

edentulous, the position of the mandible may be estimated by placing a pencil through the mandibular notch, behind the pterygoid bones and through the opposite mandibular notch. The cranium will then rest on the pencil at the approximate height above the mandible (see Fig. 6.10).

The choice of tissue depth data is determined by the sex, age and ethnic origins of the specimen (see Chapter 5). Holes are drilled into the skull at 90° to the bone surface at the appropriate number of anatomical points using a 3 mm drill bit. The holes are of varying depths, since the depth of the hole is irrelevant to the peg measurement, and the angle of the drill bit to the surface of the bone is determined by eye. Wooden pegs are cut to the lengths governed by the tissue depth data using a scalpel, a pencil and Mitutoyo 0–150 mm digital calipers, and inserted into the holes in the skull: the wooden rod is placed into the hole in the skull and the level of the surface of the bone is marked onto the rod in pencil, then the rod is removed from the hole and the appropriate tissue depth is measured from the surface-level mark on the wooden rod using the digital calipers, and marked with a pencil. The rod is then cut with a scalpel at the second mark and placed back into the hole in the skull to the level of the first mark. In this way a set of guides for tissue depth across the face is attached to the skull surface (see Fig. 6.11).

Plaster or plastic eyeballs of 25 mm diameter are set into the eye sockets, as determined by the orbital bones, using Ceramic clay (Potclays, 2003). Normal protrusion is taken as when the flat plane of the iris is touching a tangent taken from superior to inferior margins of the orbit (see Fig. 6.7). Deep protrusion is taken as the cornea touching the tangent drawn from superior to inferior margins of the orbit, and shallow protrusion is taken as the cornea being approximately 7.5 mm anterior to the same tangent. The positions of the malar tubercle and lacrimal fossa are marked on the medial and lateral orbital borders (see Fig. 4.12).

The muscles of the face are now modelled onto the skull in clay, one by one. Gross anatomy of the face and neck has already been described in detail in many anatomy books and, therefore, research regarding muscle origins, insertions and actions is referred to in Gray (1973) and Sobotta (1983). This chapter is restricted to the description of the head and neck anatomy as it relates to the shape and form of the face. Since the study of anatomy usually involves the practice of dissection, the facial musculature is normally described from the skin inwards towards the skull. However, for the process of facial reconstruction the muscles are built onto the skull one by one, from the skull outwards. Therefore, it seems more appropriate to describe the facial musculature in the direction in which the reconstruction proceeds. The first area to be reconstructed is the neck, but the majority of neck muscles are not sculpted and three main muscles are studied. These are as follows:

Sternocleidomastoid (see Fig. 6.12) is a thick rope-like muscle, which forms a prominent, visible landmark, especially when the head is turned. The *medial head* originates at the manubrium sterni and the *lateral head* originates at the upper medial third of the clavicle.

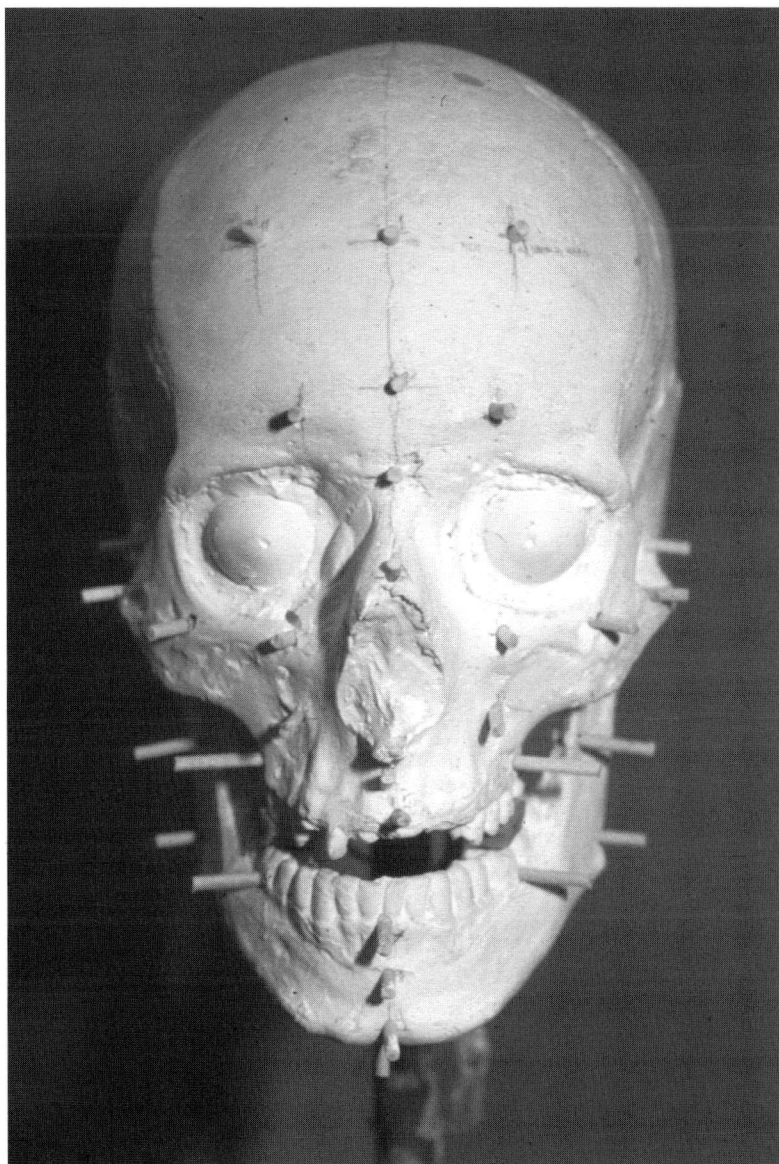

Fig. 6.11 Skull with attached facial tissue pegs. Each peg represents the mean tissue depth at that anatomical point. These data were determined by the sex, age and racial origin of the individual.

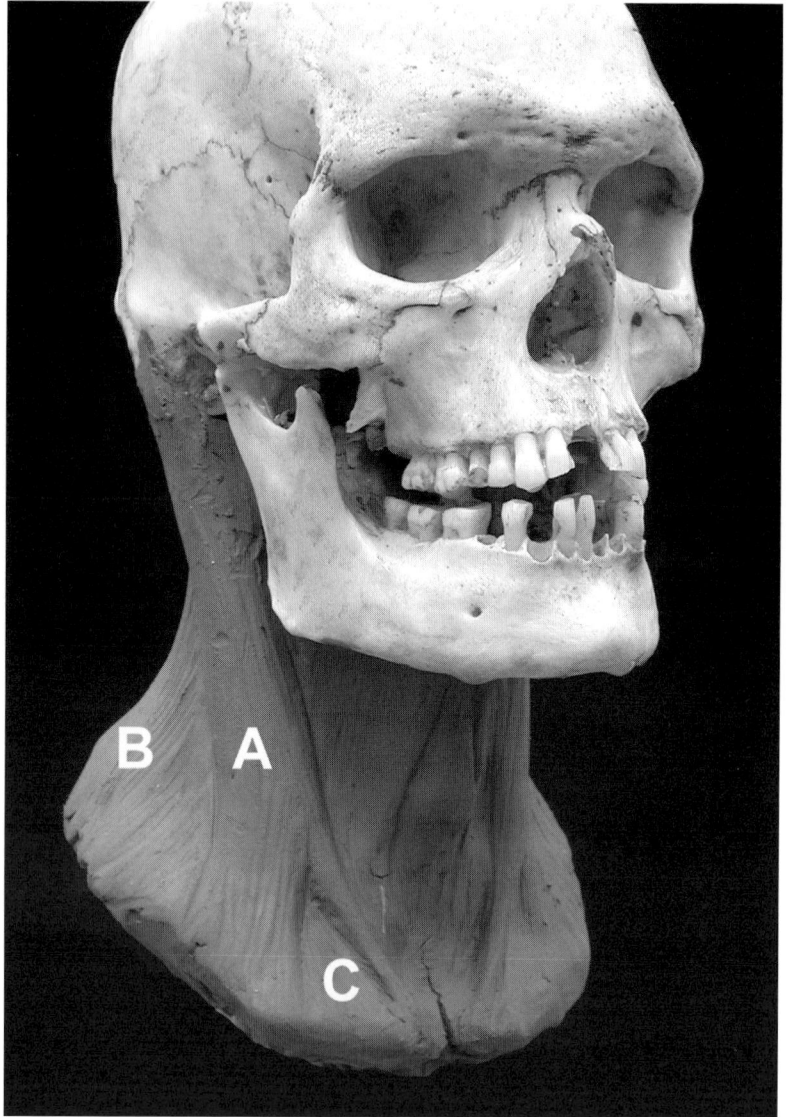

Fig. 6.12 The muscles of the neck. A = sternocleidomastoid muscle, B = trapezius muscle, C = triangular depression between the clavicular and sternal attachments of the sternocleidomastoid muscle.

The *medial head* and the *lateral head* ascend and blend with each other to form a rounded belly, and this muscle divides the side of the neck into two triangles and attaches at the mastoid process. The origins of the tendons at the clavicle and the sternum create the appearance of a small sunken triangle above the superior border of the clavicle. This muscle can be attached as a large sausage shape representing the muscle bulk, and then the attachments and origins can be sculpted. In profile, the *sternocleidomastoid* muscle is almost vertical, and from a frontal view it slopes medially downwards.

Platysma (see Fig. 6.12) is a broad, sheet muscle, which covers the clavicle, neck and mandible.

Trapezius (see Fig. 6.12) is a broad, sheet muscle, which divides the neck diagonally, almost in parallel with the *sternocleidomastoid* muscle.

The muscles of the neck are sculpted as a main block, making sure that the neck is wide enough to support the head, that the forms of the throat such as the thyroid cartilage, trachea, laryngeal prominence (or Adam's apple) in men and suprasternal notch are included, and that the curve of the neck onto the back is not too upright. An articulated stand may be employed to support the neck and shoulders, and the extent to which the shoulders are modelled is down to personal choice. Certainly, in forensic cases, the aim is to achieve as high a degree of accuracy as possible, and, therefore, shoulders must only be attempted when skeletal measurements are available.

The muscles of the face and scalp are modelled for the production of the reconstruction, in the following order:

Temporalis (see Fig. 6.13), is a fan-shaped muscle situated at the side of the head. The muscle fibres originate from the inferior temporal line, pass behind the zygomatic arch and insert at the coronoid process and anterior surface of the ramus of the mandible. The muscle fills the space behind the zygomatic arch and becomes progressively thinner as it reaches the surface of the skull. From a frontal view this muscle gives the side of the head a smooth rounded appearance. Where available the tissue depth pegs can be used as guides to the bulk of this muscle, and a couple of millimetres of peg should be left visible above the muscle surface.

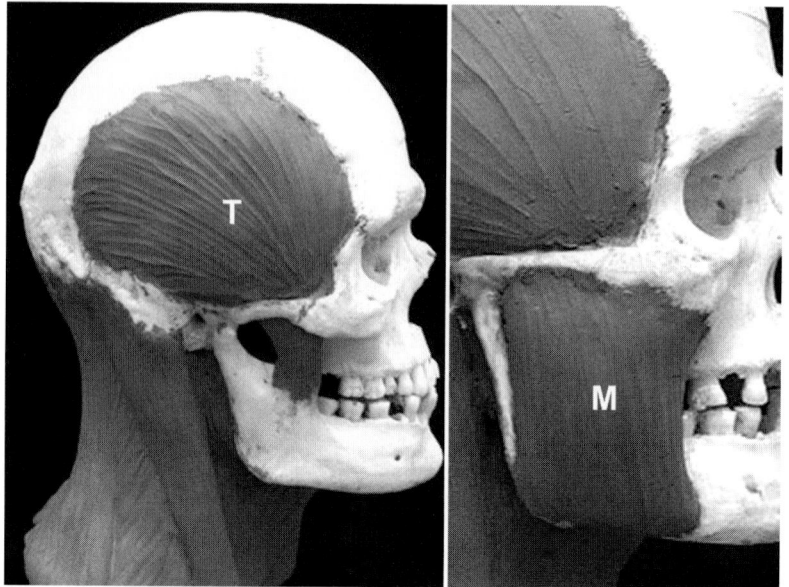

Fig. 6.13 Temporalis (T) and Masseter (M) muscles.

Masseter (see Fig. 6.13) is a bulky, strong muscle, which fills out the cheek at the side of the jaw. The muscle is rectangular in shape and from a frontal view the muscle curves gently from the inferior zygomatic arch border and zygomatic process of the maxilla to the inferior edge of the ramus of the mandible. The tissue depth pegs on the ramus of the mandible can be used as guides to the size of this muscle, and a couple of millimetres of peg should be left visible above the muscle surface.

Buccinator (see Fig. 6.14) is a quadrilateral muscle, originating from behind the mandibular ramus at the alveolar processes of the maxillary and mandibular molars, partially filling the space between *masseter* and the teeth. *Buccinator* fibres converge towards the angle of the mouth where they become continuous with the fibres of the *orbicularis oris* muscle and the lips. This muscle is known as the 'trumpeter' muscle as it compresses the cheek against the teeth.

Orbicularis oris (see Fig. 6.14) is a circular muscle, which covers the teeth and forms the slit of the mouth and the basis of the lips. Experience suggests that the most effective way to model this

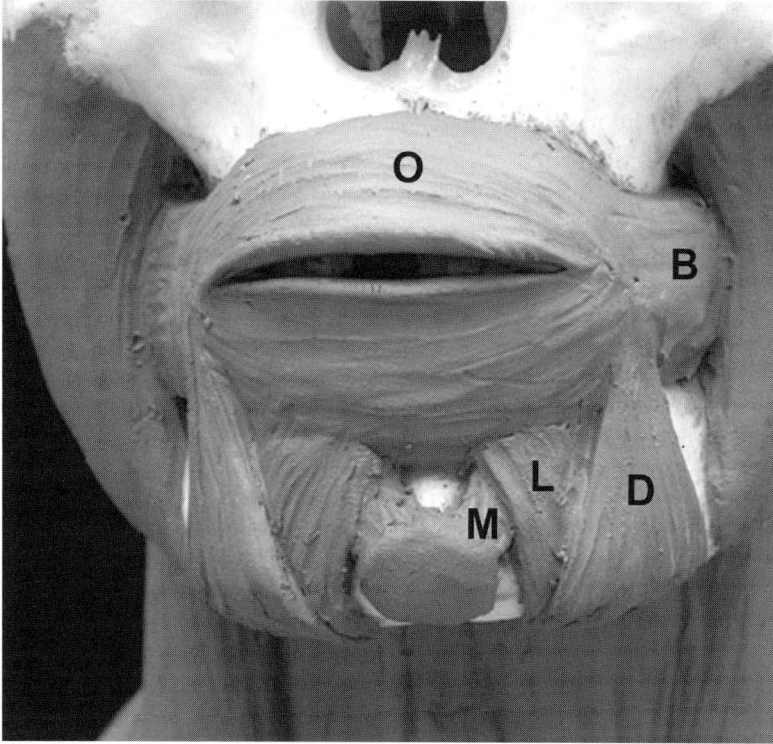

Fig. 6.14 The muscles of the mouth and chin. O = orbicularis oris, B = buccinator, M = mentalis, L = depressor labii inferioris, D = depressor anguli oris muscles.

muscle is to roll a sausage shape, and position it by bending it across the teeth in an oval. The sausage can then be flattened, whilst positioning the corners of the mouth on a radiating line from the lateral border of the canine teeth, and the mouth fissure at the middle of the upper incisors. The *orbicularis* oris should be modelled as a thick muscle, using the tissue pegs as guides, as the clay will represent the space between the teeth and the soft tissue, as well as the muscle. A couple of millimetres of peg should be left visible above the muscle surface. The mouth may be modelled open or closed, dependent upon the dental occlusion of the individual. When the individual has prominent maxillary canines, maxillary prognathism, or an overjet, it will be appropriate to model the mouth slightly open, with the maxillary teeth visible, as this is the characteristic relaxed position for these types of occlusion. Distinctive or unusual dental detail, such as diastemas, missing teeth or ornate dental work, should be made visible by modelling the mouth open, as these details could be significant in the facial recognition. It is inappropriate and unnecessary to model a forensic facial reconstruction smiling. The majority of dental information can be visible with an open, but unsmiling, mouth. The *orbicularis oris* muscle has substantial bulk and forms a 'J' shape in profile, where the lips attach to the muscle. The 'J' shapes can be modelled by rolling two thin sausages, and attaching them above and below the mouth fissure, onto the surface of the muscle. The lips should not be modelled at this stage, as their shape and form will be affected by the muscles of the cheek. Lip sculpture is usually reserved until after the skin layer has been applied. Numerous muscles converge at the angle of the mouth and they interlace at a palpable nodular mass known as the *modiolus*, where the muscle fibre direction appears unclear.

Mentalis (see Fig. 6.14) arises from the incisive fossa of the mandible and inserts at the tissue pad of the chin. The fibres form a conical muscle, which lies below the depressor muscles.

Depressor labii inferioris (see Fig. 6.14) is a quadrilateral muscle, which passes from the inferior border of the mandible to the skin of the lower lip. This muscle forms the lateral chin shape, and overlaps *mentalis* at right angles to the direction of the *mentalis* muscle fibres.

Depressor anguli oris (see Fig. 6.14) is a fan-shaped muscle, which originates at the inferior border of the mandible and inserts at the *modiolus* to create the lateral jawline. It overlaps *depressor labii inferioris* and is the most superficial of the muscles of the chin. Where available, the tissue depth pegs can be used as guides to the bulk of this muscle, and a couple of millimetres of peg should be left visible above the muscle surface.

Orbicularis oculi is a broad, flat, circular muscle (see Fig. 6.15), which occupies the eyelids, surrounds the orbit, and spreads up over the temporal region and down over the zygomatic process of the

Fig. 6.15 Orbicularis oculi muscles.

maxilla. The edges of this muscle are determined by the shape
of the orbit, brow and maxilla. From experience, the best way to
model this muscle is to roll a sausage shape and position it around
the orbit, leaving a gap between the sausage and the eyeball. The
sausage can now be flattened onto the bone around the orbit, and
smoothed so that it follows the shape of the orbital rim and bends
into the orbit. The circumference can be trimmed so that it is
not more superior than eyebrow level, not more inferior than the
canine fossa, not more medial than the internasal suture, and not
more lateral than the lateral border of the zygomatic bone. Thin
strips of clay can be placed over the eyeball to represent the eyelids, and the positions of the inner and outer canthi are determined
from the positions of the lacrimal crest and malar tubercle respectively. The eyelids should just intersect the border of the iris and
should hug the eyeball, following its shape. At the canthi, the lids
leave the surface of the eyeball, and meet at the caruncle of the
inner canthus and the angle of the outer canthus. The total lower
lid shape forms an 's' curve. The upper lid is superficial to the lower
lid at the outer canthus. The eyes of Mongoloid faces do not expose

all of the lacrimal caruncle at the inner canthus, and the line of the upper lid will be straighter, leaving the surface of the eyeball earlier than in Caucasoid and Negroid faces (Wen, 1934).

The basic nasal shape is then modelled with the nasal projection taken as the point at which the tangents from the last part of the nasal root bones and the nasal spine intersect (see Fig. 6.16). The nasal cartilages are modelled onto the nasal aperture (see Fig. 6.17) following the line of the nasal bones. The cartilage shape does not provide the finished nasal shape but will suggest the profile and alae position. The alar shapes are modelled with reference to the nasal aperture contour and nasal bone form (see Fig. 4.9). Two balls of clay are rolled, with diameters roughly equal to one-fifth of the overall width of the nose, and placed onto the sides of the nasal cartilage. These balls should be placed approximately 4 mm lower than the inferior border of the nasal aperture, leaving space posterior to the balls and anterior to the muscles, where further soft tissue will be added to the cheeks. The balls of clay can now be modelled to represent the appropriate alar and nostril shape and the nasal tip can also be modelled at this stage.

Levator labii superioris alaeque nasi (see Fig. 6.18) is a thin strap-like muscle, which runs down the side of the nose from the lateral nasal bones. A medial branch attaches at the alae and nasal skin, and a lateral branch attaches at the upper lip.

Nasalis is a thin strap-like muscle (see Fig. 6.18), which runs down the side of the nose from the bridge of the nose. There is a small medial branch that attaches to the alae, and a larger lateral branch that attaches to the maxilla, lateral to the nasal notch.

Levator anguli oris muscle (see Fig. 6.18) has its origin at the canine fossa and infraorbital margin and converges with the other muscles at the *modiolus*. *Levator anguli oris* is a thin flat muscle, which lies deep to the *levator labii superioris* and *zygomaticus minor* muscles.

Levator labii superioris muscle (see Fig. 6.18) is a thin muscle that appears to gently fan out from the upper lip to the lower border of the orbit and zygomatic bone. This muscle lies slightly

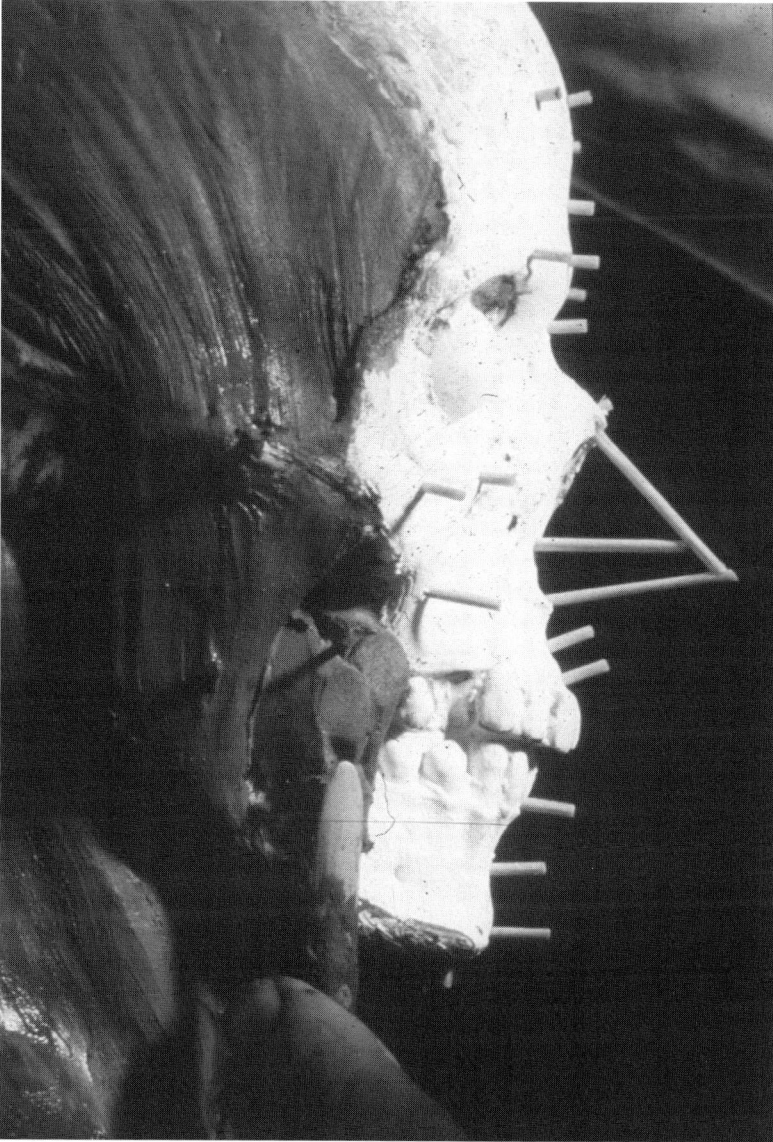

Fig. 6.16 The nasal projection. Tangents from the last part of the nasal bone and the nasal spine will determine the nasal projection.

above the *levator anguli oris* muscle, although the muscle fibres of these two muscles and the fibres at the medial end of the *zygomaticus minor* muscle can appear indistinguishable in life. The two *levator* muscles can be modelled as one block, which should follow the contour of the canine fossa.

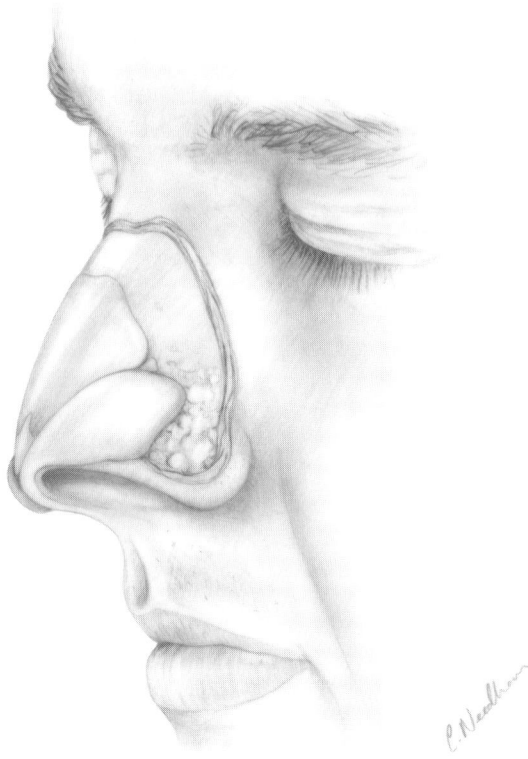

Fig. 6.17 The nasal cartilage shape.

Zygomaticus major and *minor* muscles (see Fig. 6.19) are strap-like muscles that form the contour of the cheek. The *zygomaticus minor* muscle arises from the lateral surface of the zygomatic bone, immediately lateral to the zygomaticomaxillary suture, and passes downwards and medially to insert into the upper lip. The *zygomaticus major* muscle arises from the zygomatic arch, anterior to the zygotemporal suture, and inserts into the angle of the mouth where it converges with all the other muscles at this point. When the form of the cheekbones is flat, the zygomatic muscles are located more frontally, and when the cheekbones are strongly profiled, the zygomatic muscles are located more laterally. These muscles can be modelled as thin strips of clay, but it is sensible to fill the space behind these muscles to act as support. When supporting clay is not attached behind the zygomatic muscles, the muscles

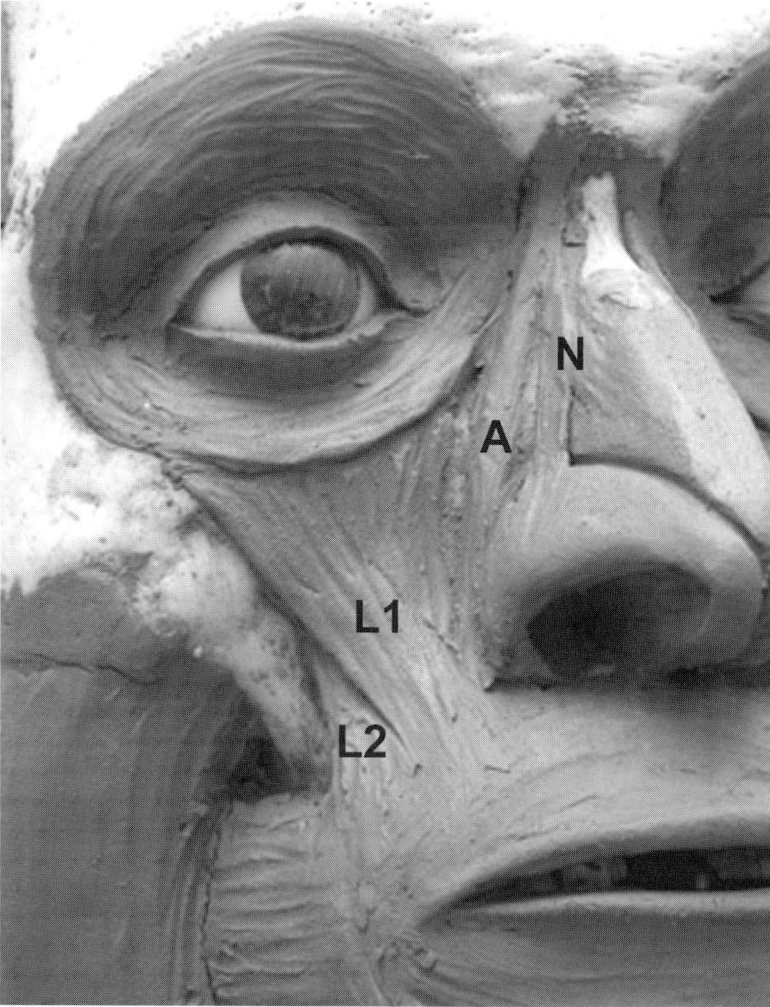

Fig. 6.18 Nasal muscles. A = levator labii superioris alaeque nasi, N = nasalis, L1 = levator labii superioris, L2 = levator anguli oris muscles.

will sometimes collapse under the pressure of the skin layer application, and will suggest a distorted cheek contour. Where available the tissue depth pegs can act as guides, and a couple of millimetres of peg should be left visible above the muscle surface. It is usually at this stage that the reconstruction practitioner may wish to deviate from the tissue depths suggested by the pegs. The tissue depth pegs represent average tissue measurements appropriate to

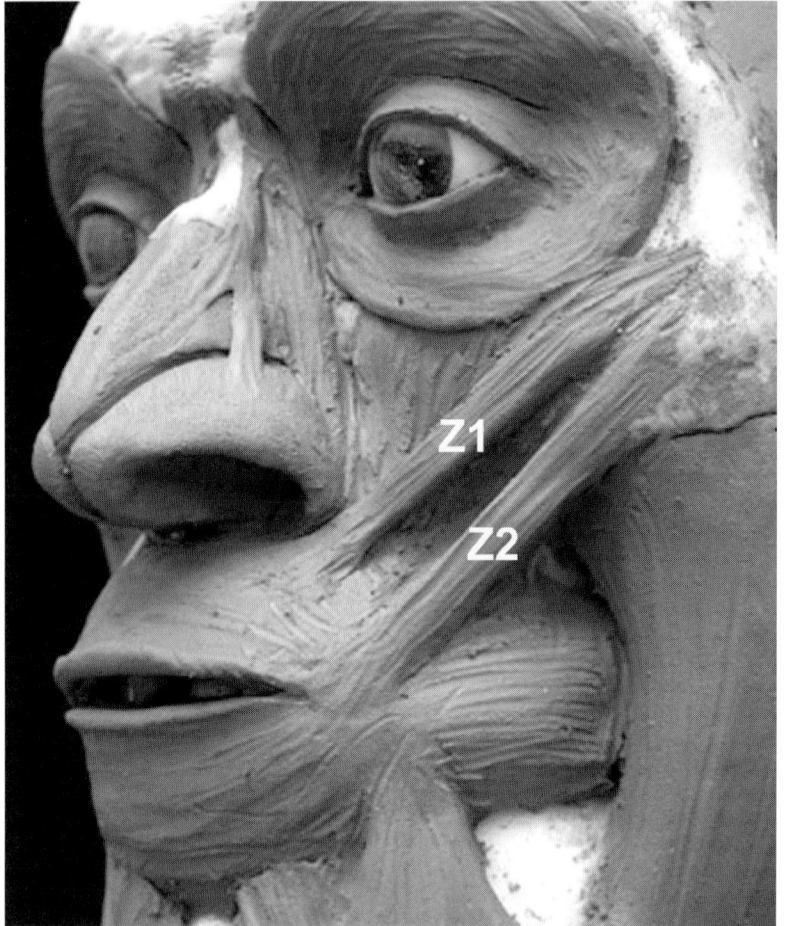

Fig. 6.19 The muscles of the cheek. Z1 = zygomaticus minor, Z2 = zygomaticus major.

the age, sex and racial origin of the skull, but since these are averages, they will not always be accurate for an individual skull. Frequently, the skull morphology will suggest that one or more of the tissue pegs are misleading, and the reconstructor should always follow the facial contour suggested by the skull. Where the pegs are too projecting or too shallow they can be removed or ignored. The facial musculature will determine the face shape and contours and the tissue depth pegs should only be considered as guides.

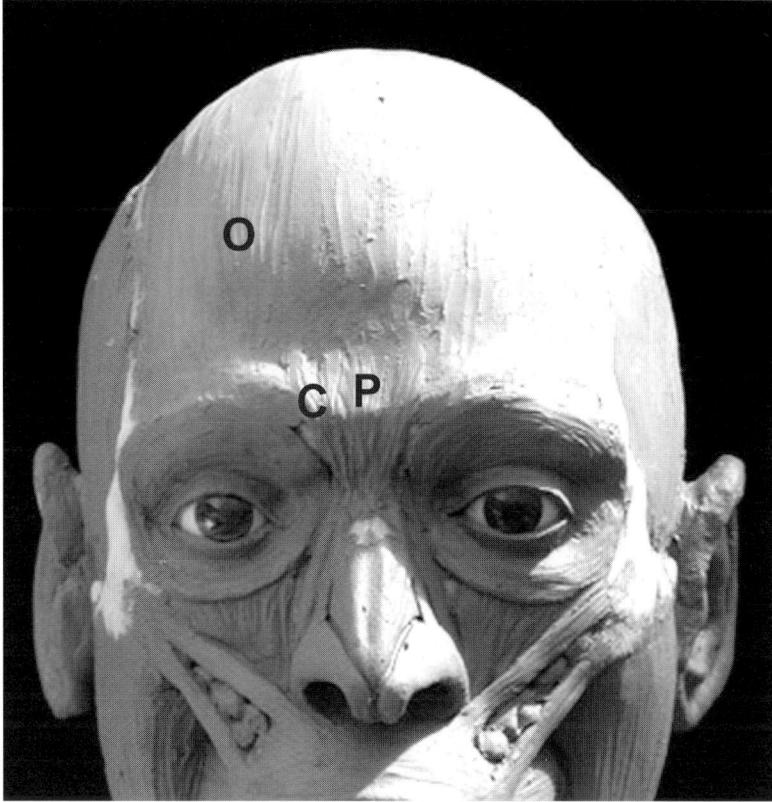

Fig. 6.20 The muscles of the brow. O = occipitofrontalis, P = procerus, C = corrugator supercilii muscles.

Corrugator supercilii is a small pyramidal muscle (see Fig. 6.20) at the medial end of the eyebrow. This muscle is not strictly necessary for the reconstruction procedure, nonetheless it is often included.

Procerus (see Fig. 6.20) is a triangular muscle of facial expression and is not strictly necessary for the facial reconstruction procedure, although again it is often included. The origins are at the lower nasal bone and upper lateral nasal bone and inserts at the skin of the forehead between the eyebrows.

Occipitofrontalis (see Fig. 6.20) is a thin, broad, musculofibrous layer, which covers the top of the cranium from the nuchal lines of the

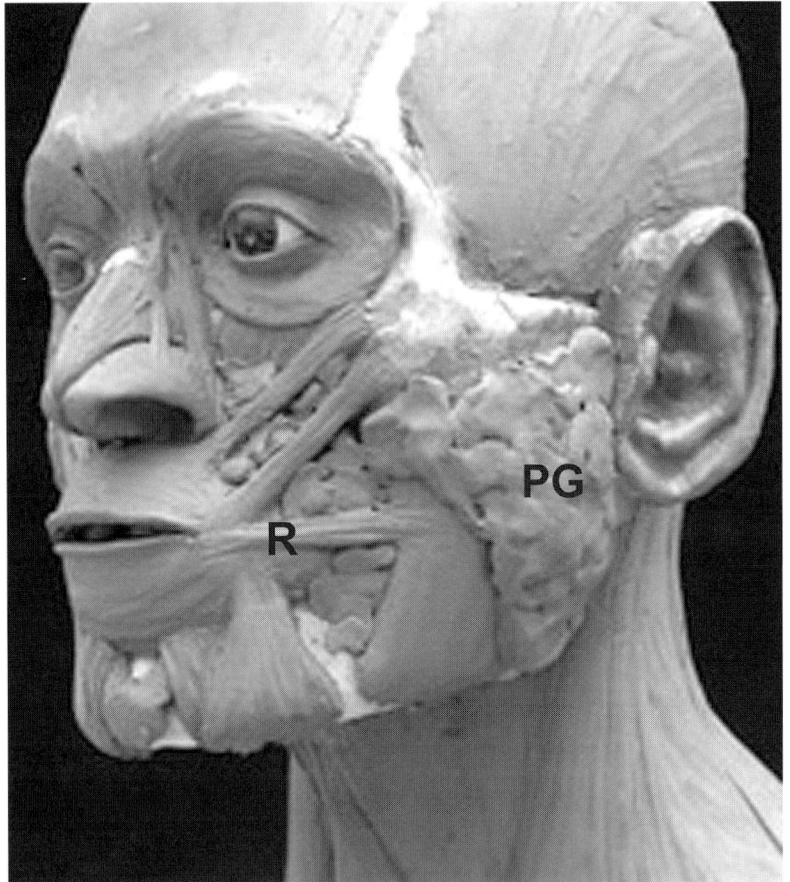

Fig. 6.21 The parotid gland (PG) and risorius (R) muscle.

eyebrows. Where available the tissue depth pegs can be used as guides to the bulk of this muscle, and a couple of millimetres of peg should be left visible above the muscle surface.

The *parotid gland* is then modelled onto the face (see Fig. 6.21). The parotid gland is an irregular, lobulated mass, lying below the external auditory meatus, between the mandible and the *sternocleidomastoid* muscle. It projects forward onto the surface of the *masseter* and a small part of it lies below the zygomatic arch and above the parotid duct. This gland is modelled using small lumps of clay that are applied to the surface of *masseter* in a random pattern within the boundaries of the gland shape. Smoothing of the surface with a wet sponge will create a gland-like appearance.

Risorius is a strap-like muscle with origins at the parotid fascia and insertions at the corners of the mouth (see Fig. 6.21).

The ears are modelled with reference to the mastoid processes and angle of the jaw, and these are attached above the external auditory meatus on a block of clay, which represents the skin thickness. The ears are notoriously difficult to model with any degree of accuracy, and there is much detail regarding the shape, projection and size that cannot be determined from the skeletal morphology. Some practitioners (Wilkinson *et al.*, 2002) use ear casts for forensic investigations, attaching a pair of small, medium or large ears dependent upon the size suggested by the mastoid processes, nose length and the head size. The ear position, angle and projection can then be altered for each reconstruction and, where necessary, the lobes can be altered. This requires sets of ear moulds taken from life, but can be a very time-effective method. Other practitioners (Prag & Neave, 1997; Taylor, 2001) suggest modelling the ears, and Taylor describes a very useful step-by-step method for ear sculpture. In some forensic, and archaeological, investigations the soft tissue of the ears may be undamaged (for example, partial decomposition, water submergence, peat bog bodies or Egyptian mummies). Obviously in these cases it is preferable to sculpt the ears from the available detail. The position of the external auditory meatus suggests the position of the external ear hole, and the ears are attached using the jawline and nasal profile as guides (see Fig. 4.17). The ears must be attached on a block of clay so that they are raised off the surface of the muscles at a distance equal to the soft tissue layer to be added. This temporarily gives the face a simian appearance, which will disappear once the skin layer is applied.

The resulting facial musculature will already illustrate the basic face shape and proportions (see Fig. 6.22). Different skulls will produce different musculature shapes, and it is at this stage that

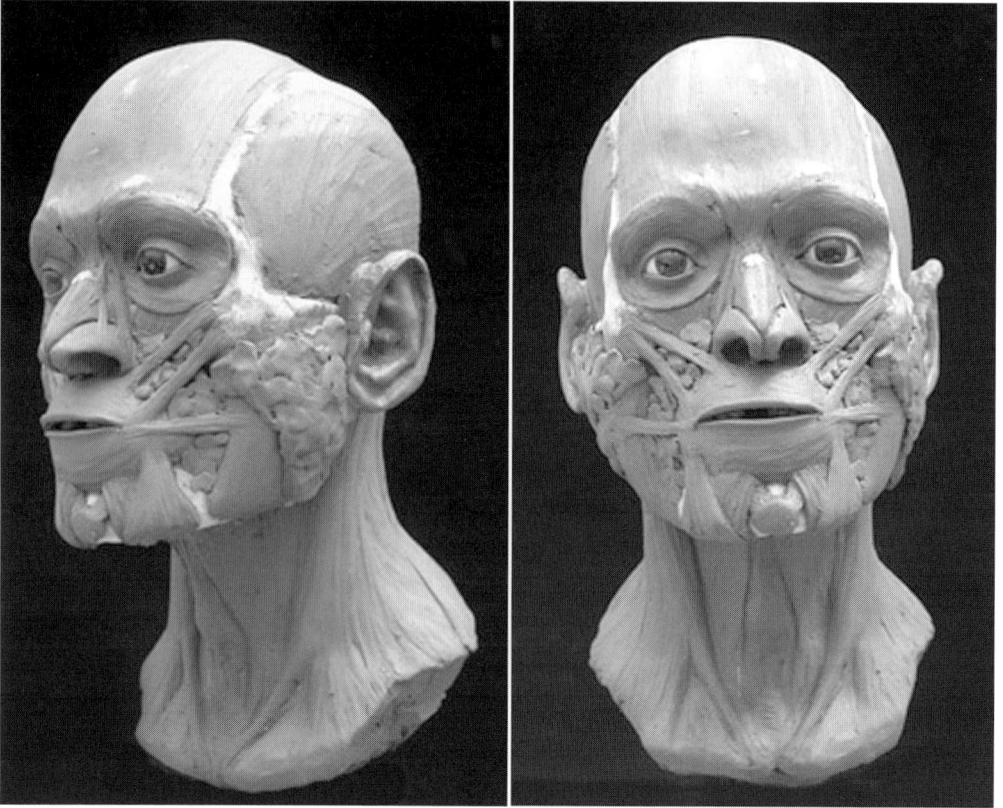

Fig. 6.22 The reconstruction of the facial musculature.

it becomes clear how much the skull directly influences the facial appearance (see Fig. 6.23). As Neave stated, '*There are those who argue that to model the underlying structures in such detail is unnecessary... however, this approach is the most logical and foolproof way of ensuring that the face grows from the surface of the skull outwards of its own accord... and reduces to a minimum the possibility of subjective interference by the artists.*'

The next stage is to place a layer of clay over the musculature that represents the skin and subcutaneous fat. It is helpful to place small pieces of clay between and over some of the muscles to represent the subcutaneous fat (see Fig. 6.22) at the cheeks, prior to application of the skin layer. This helps support the skin layer and stops hollows from forming between some the muscles, such as

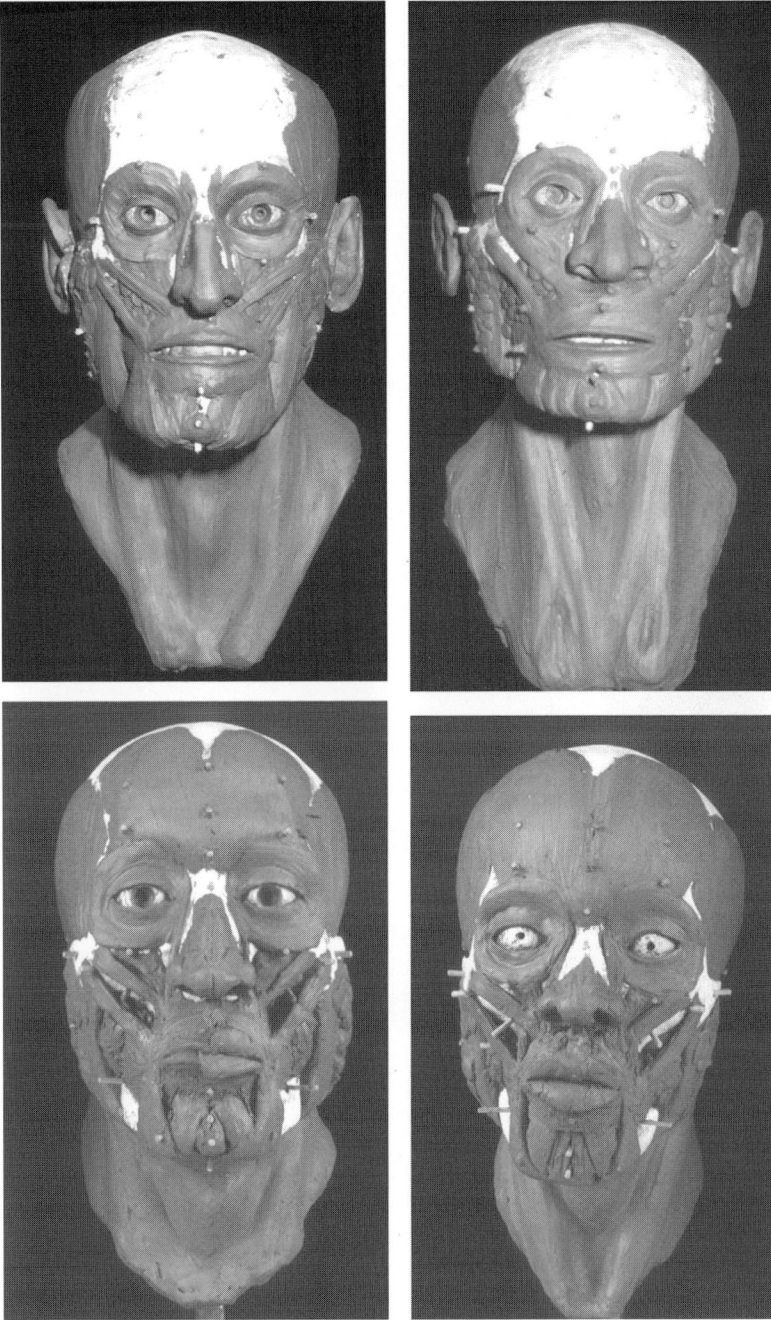

Fig. 6.23 Different reconstructions at the muscle stage. The different proportions and form of the facial musculature are a direct result of the variation of the skull.

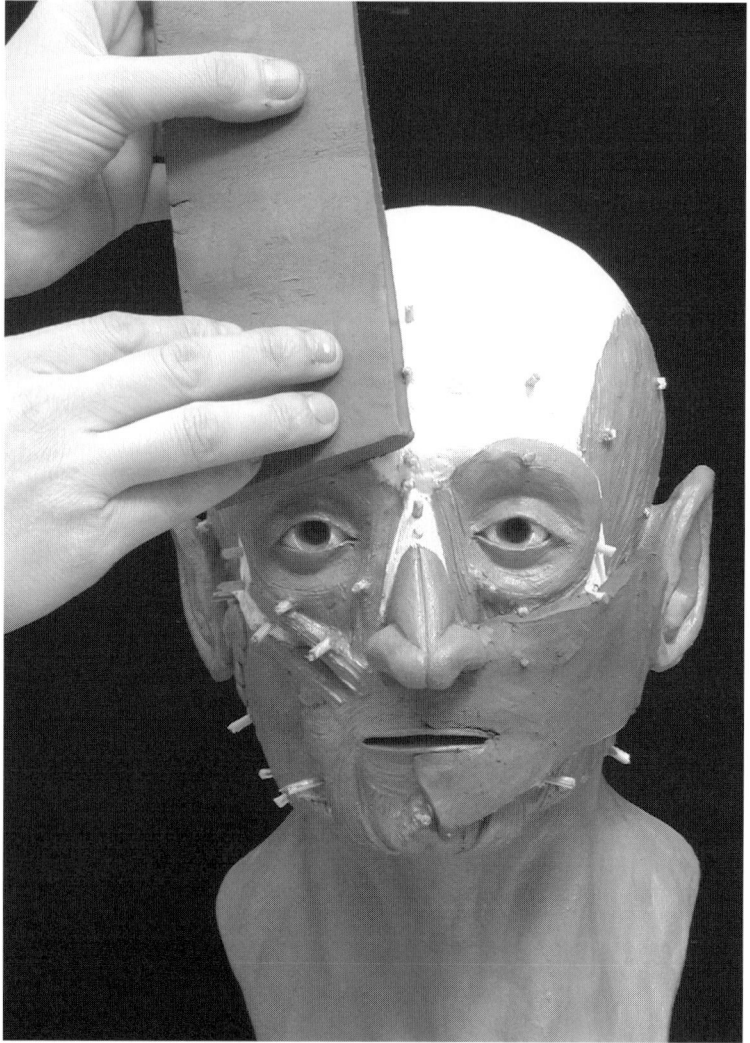

Fig. 6.24 The addition of the skin layer.

the *zygomatic* muscles. Strips of clay are then rolled, shaped, and placed over the muscle/fat structure to create the finished face (see Fig. 6.24). The thickness of the skin layer is determined by eye, as approximately a third of an inch (8.5 mm), but the tissue guides over the face determine the actual skin thickness, and extra clay may need to be added or removed at some areas. At the half-muscle, half-skin stage it is evident that the skin layer has followed the underlying muscle structure, and the two sides of the face are basically the same shape and proportions (see Fig. 6.25). The same relationship can be seen when the opposite procedure is followed, and the face is dissected back from the skin to the muscles. The details of the facial features (nasal shape, lip form and eyebrow pattern) are modelled with respect to the assessment of the skull, the details of which have been recorded earlier. The surface of the face is smoothed and a final sculptural finish achieved (see Fig. 6.26). Students often find this part of the reconstruction procedure the most difficult, and a reconstruction that appears accurate and well formed at the muscle stage can become wooden and mask-like following skin application. A practitioner with poor sculptural skills will have difficulty producing a realistic and believable face, and an artist who does not rigorously follow scientific rules will have difficulty producing an accurate reconstruction.

In order to preserve integrity, every detail of the reconstructed face must be based on scientific assessment of the skull morphology. The nose must be modelled following the morphology suggested by the nasal bones, brow and cheekbones, the mouth must be developed from the dental pattern, cheek muscles and palate shape, the eyes must be sculpted from information supplied by the brow, orbital bones and nasal root. The skull must be studied to determine the presence or absence, position and strength of the nasolabial fold, mental crease, vertical glabellar lines, nasoglabellar line, buccal crease and forehead creases. Facial morphology that cannot be determined from the skeletal detail with any certainty, such as the nasal tip, must be estimated in sympathy with the rest of the face. This is really dependent upon experience. In forensic cases, hairstyles, facial hair, wrinkles, blemishes, scars and glasses should only be added to the reconstruction where these details are suggested by the human remains. False information may confuse the observer and obscure recognition. Age-related facial details must be considered if appropriate, such as eye bags, neck sagging, jawline softness and drooping of the eyelid. Facial appearance details may help to suggest the age of the individual, but care must be taken to

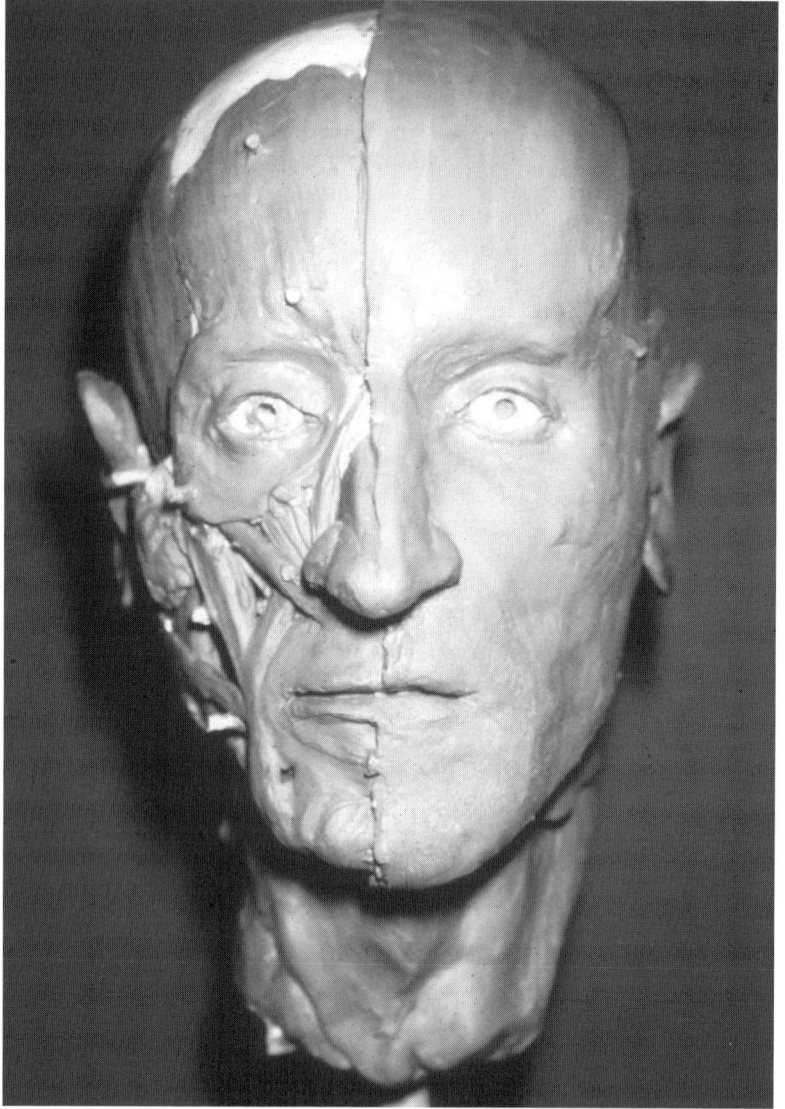

Fig. 6.25 A half-muscle/half-skin reconstruction. The facial proportions and shape are the same on both sides of the face. The skin layer has just padded out the facial musculature and it is the skull shape that directly influences the facial form.

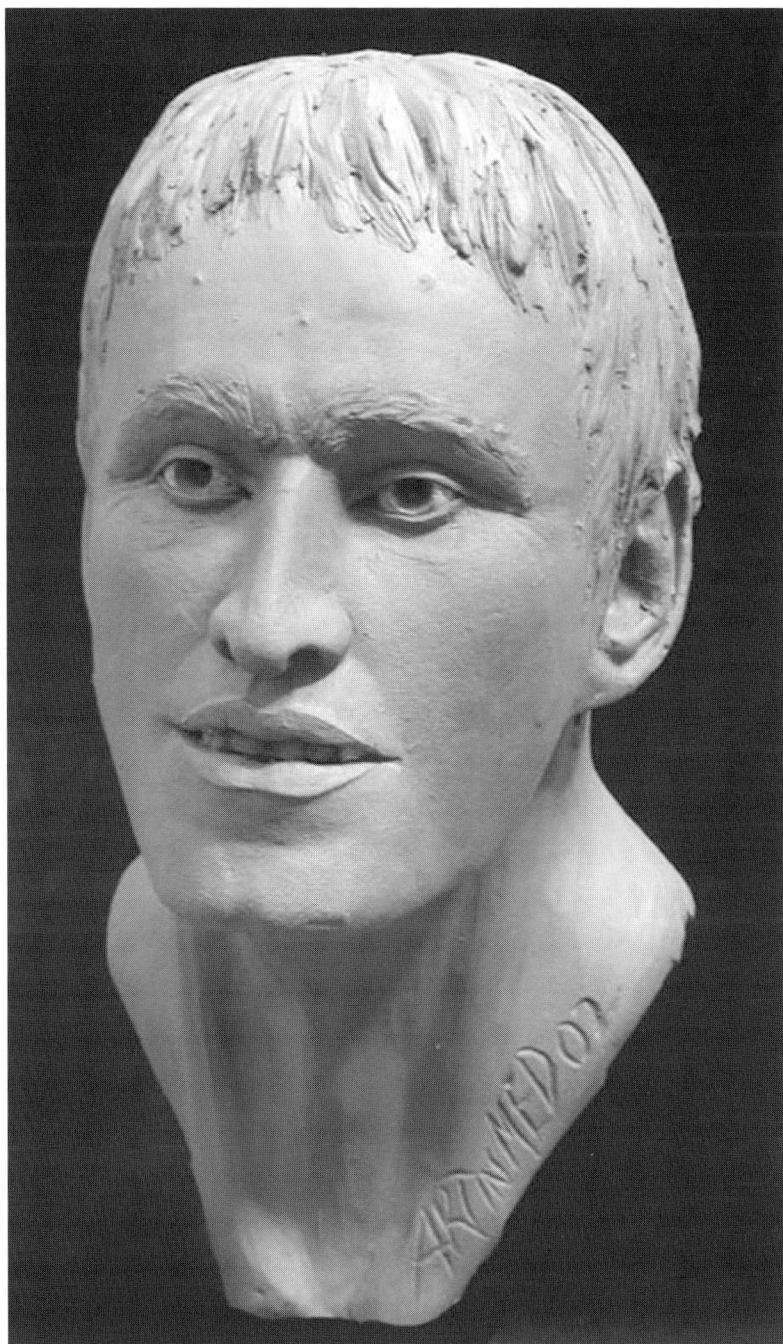

Fig. 6.26 An example of a finished facial reconstruction.

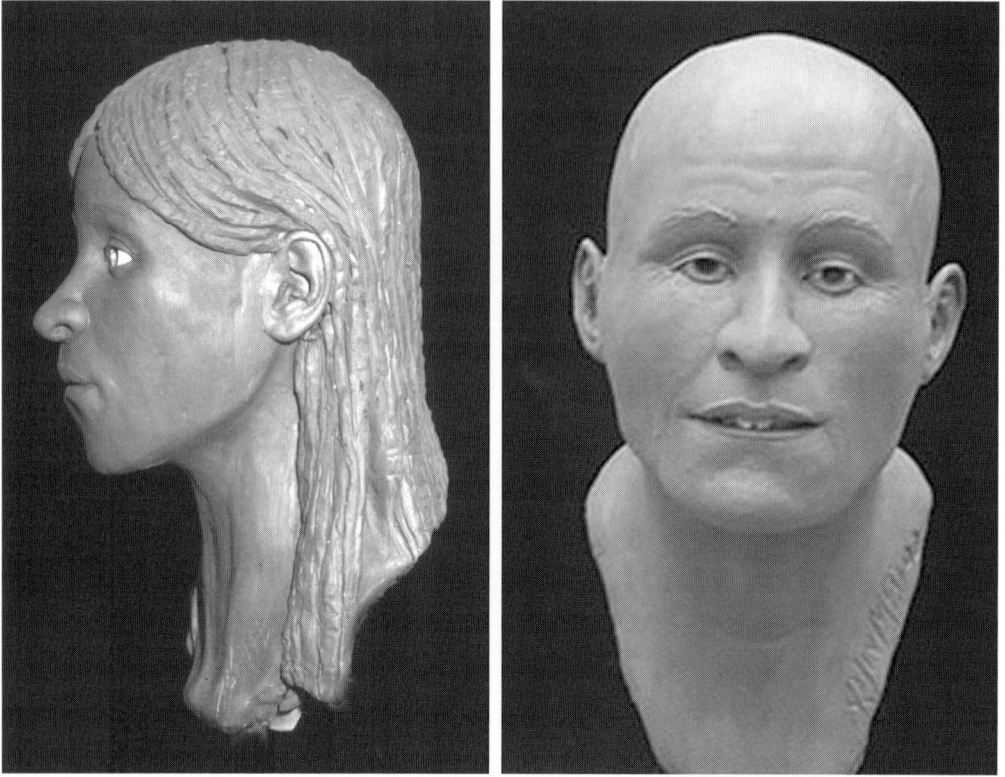

Fig. 6.27 Qurneh mummy (left) and priest (right). These mummies were from Ancient Egypt and were given hairstyles as suggested by written text and appropriate to their standing, age and period of life. The Qurneh mummy (1570 BC) had an ornate coffin and was thought to be someone very important, such as a queen or princess. She was a young adult and was given a complicated wig made from many thin plaits. Iufemamun was over 40 years of age and was known to have a shaved head, as was common for priests of this period (1085–945 BC). These reconstructions can be seen at The National Museum of Scotland, Edinburgh.

avoid too noticeable aging, as these details can only be estimates, and false impressions of appearance may be produced. In contrast, the American method, as described by Taylor (2001), will estimate eye colour, skin tone etc. from population statistics. For example, Taylor states that,

> *eye colour choice for skulls designated by the anthropologist as being Negroid or Mongoloid would most logically be reconstructed with brown prosthetic eyes. When light blond hair specimens are recovered, a lighter eye choice may be suggested. Hazel eyes are a good all-round choice for*

Caucasoid skulls. Obviously any eye colour can occur in any group, but eye colour choices are best made based on the odds.

Practitioners of the Manchester method initially prefer to add only those facial appearance details that are certain. The wrong eye colour, hair colour, skin tone etc. may lead to the reconstruction not being identified. It is possible that an observer may think he recognises the individual, but then thinks, '*Oh, it can't be Uncle Bob because he had blue eyes.*' However, in some forensic cases the initial facial reconstruction may not elicit a response from the general public that leads to identification, and in these cases further images of the reconstruction may be produced depicting hairstyles, skin colour, eye colour etc. These images are often produced by the addition of colour, hair etc. to an image of the reconstruction, using computer software such as Adobe Photoshop. The images may provide additional estimation of facial detail and, although they may elicit a further response from the general public, it must be noted that the images include a great amount of un-certain detail. In archaeological investigations, this caution is not necessary, and hairstyles, eye colour, hair colour, skin tone etc. will frequently be estimated by the archaeologists or Egyptologists. Hairstyles appropriate to the period of time and the class, sex and age of the individual will be modelled onto the reconstruction or added to the wax head in real hair (see Fig. 6.27).

7   The accuracy of facial reconstruction

The accuracy of the facial reconstruction method has been extensively debated over the years. One of the points of contention seems
to be disagreement over who should carry out the facial reconstruction work itself. Should it be the artist, who uses more intu-
itive sculptural skills; or the scientist, who follows a preconceived
method? Often scientists employed artists to carry out the sculptural work under their scientific supervision. Such work, however,
did not escape criticism. As early as 1900, Merkel was assisted by
Eichler, a sculptor, and together they carried out the reconstruction
of an ancient Saxon. Eichler departed widely from the prescribed
data and followed his own deductions concerning the development
of muscles, as suggested by the lines and features of the skull.
The aesthetically pleasing result could not be tested and therefore
the departure from the data could not be defended. Alternatively,
there have been many facial reconstructions that have been carried
out with a very methodical and scientifically pure technique that
have resulted in wooden, lifeless and clumsily sculpted faces that
are crying out for the aesthetic and observed input of an artist.
The contention is based around whether or not aesthetic appeal
is a necessity when recognition or similarity to an individual is
the only requirement. One theory proposed by George (1993) suggested that the face shape and proportions are the most important
factors, and any added detail, which makes the face human, may
detract from the features and overall impression of the face. However, some studies suggested that the face has to appear human and
alive to spark the recognition required for forensic identification,
and that badly sculpted faces may detract from any recognition
(Suk, 1935; Bartlett et al., 1984). Certainly, with archaeological facial
reconstruction, the merits of good sculpture and lifelike, believable

faces are obvious. Some scientists have carried out facial reconstructions with the fine skills of a practised artist (Gerasimov, 1971; Helmer, 1984) and there have been reconstruction artists with medical training (Neave, 1979; and Taylor, 2001) who have followed the scientific procedures whilst allowing their artistic skills to interpret the data. Either variation seems to have produced the optimal investigator for such reconstruction work. However, there are many current practitioners of facial reconstruction who have neither the artistic skills nor the scientific rigour, and it is these practitioners who continue to give this field a bad name.

Early attempts using cadavers

Over the last 120 years there have been many studies into the reliability and accuracy of the reconstruction techniques. The early attempts at the reconstruction of Prehistoric Man did not offer much hope for the accuracy of facial reconstruction. Several different anthropologists from across the world reconstructed the Neanderthal skull from La Chapelle-aux-Saints, France in 1908, and the results differed markedly from each other. However, research into the accuracy of facial reconstruction required comparison of the resulting face with a realistic representation of the individual. The first studies were compared with busts and paintings of historical figures, such as Bach (His, 1895) and Dante (Kollman, 1898), and the results were promising (see Fig. 2.5). However, the sculptors already had knowledge of the appearance of these historical figures and this could not be considered a blind study. Anatomists turned to the mortuary for subjects to study and the early accuracy studies employed death masks or photographs of the cadavers for comparison with the facial reconstruction. Von Eggeling (1913) produced the first recorded accuracy research. He made a death mask from a corpse and measured soft tissue on the face at various points. Two plaster casts of the skull were made and given to two sculptors, together with the soft tissue data. The facial reconstructions were carried out independently and were found to have no resemblance

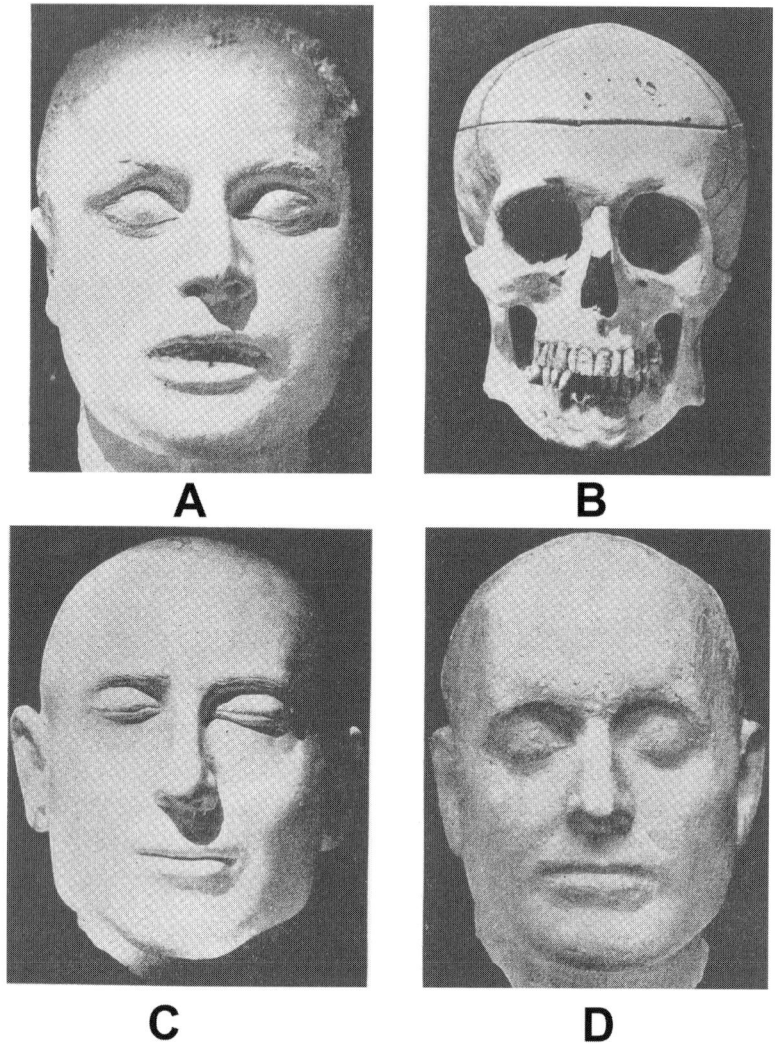

Fig. 7.1 Facial reconstruction accuracy test by Von Eggeling (1913). A = death mask of subject, B = skull of subject, C = reconstruction 1 by Bergemann-Konitzer, D = reconstruction 2 by Elster. After: *Personal Identification* (Wilder & Wentworth, 1918).

to one another nor to the individual (see Fig. 7.1). However, Gerasimov attributed this failure to the '*carelessness*' of the two sculptors who '*did not pay great attention to the correlation between the shape of the skull and the thickness of the soft parts*'. Wilder (1912) was a great proponent of the facial reconstruction procedure and stated that the method is so simple that it can be readily performed by any-one who follows the directions, and the very first attempt cannot help being at least moderately successful. However, he did warn that any weakness lay in the opportunity for imagination on the part of the investigator at such features as the lips, the soft parts of the nose and the set of the eyes. Stadtmuller (1922, 1923) car-ried out the next reported study. Stadtmuller and his assistant, Zimmerman, both reconstructed the skulls of a 62-year-old man and a 17-year-old man, using the Kollman and Buchly (1899) soft tissue data. Each attempt was compared with photographs of the corpse and with death masks. No similarities were observed be-tween the reconstructions and the individuals.

They then carried out the same reconstructions again with the aid of photographs of the deceased. Each reconstructor was forced to deviate from the soft tissue data in order to make the faces 'fit' the photographs. Stadtmuller then stated that such deviation called into question the validity of average, as opposed to individ-ual, values for soft tissue thicknesses. A few years later, in 1926, Diedrich made yet another study on five corpses, checking the re-sults against traced photographs of the cadavers. Again, none of the reconstructions was recognisable as portraits of the deceased indi-viduals. Diedrich therefore concluded that a reconstruction pro-vides only an approximation of a basic head type.

Suk (1935), a Czech scientist, suggested that it was a great mis-take to consider the features of the face to be dependent on the bony structure of the skull and concluded that a facial reconstruc-tion from the skull must resort to fantasy. Brues (1958) agreed with Suk and, as said earlier, stated that facial reconstruction is '*proba-bly best left to the ample literature of detective fiction*'. However, not all cadaver comparison studies were so disheartening. The facial anthropologist, Gerasimov (1971), working in Russia, was not deterred by the negative results of Stadtmuller and Von Eggeling. In 1940, he carried out a mass control experiment using 12 heads from a mortuary at Moscow Medical Institute. He produced the reconstructions without any knowledge of their actual identity and claimed that all 12 heads established a similarity with police photographs of the deceased. With characteristic humility Gerasi-mov concluded that '*the results exceeded even my expectations. All the reconstructions proved to be like portraits although they were of persons of both sexes, of various ages and of different racial affinities. All the twelve heads were so true to life that their identity with the relevant photographs was undoubted.*' In Britain, Neave (Prag & Neave, 1997) also carried out a study of four cadavers to establish the accuracy of facial reconstruction. The cadavers were photographed and Neave produced reconstructions following a rather simple and underde-veloped method based on a combination of the anatomical and tissue depth information. Neave then correctly linked the resultant four reconstructions to the photographs of the individuals, and several professional colleagues judged the reconstructions to be '*uncanny*' resemblances of the individuals depicted in the photographs. These studies made similarity assessments by comparing the reconstruction with cadaver masks or photographs, and there are obvious problems associated with the facial appearance of cadavers (see p. 129). Suk (1935) suggested that a face without life bears little resemblance to the individual in life, and this suggests that cadaver comparisons were not a wholly appropriate method of judging the accuracy of a facial reconstruction.

Qualitative studies

Fortunately, the generalised use of photography in society and the employment of facial reconstruction work in forensic identification investigations allowed further accuracy research, where the reconstructed face could be compared with a photograph of the individual during life. Gerasimov produced positive results from numerous accuracy studies and the most documented was the case of a Papuan who died from tuberculosis in 1912. Gerasimov received a great amount of contemporary criticism from leading Russian anthropologists following a lecture on his methods in 1939. In response to their mistrust of his results he suggested a blind test in the presence of his critics and that they themselves could choose

Fig. 7.2 Facial reconstruction of a male Papuan by Mikhail Gerasimov (1939). Reconstruction (left) and identified individual (right). Modified from Gerasimov (1975). Courtesy of the Laboratory of Anthropological Reconstruction, Moscow.

the skull. The next morning he was presented with a Negroid skull and Gerasimov used Kollman's anthropological data when producing the reconstruction (see Fig. 7.2). Gerasimov was not provided with any information regarding the racial origins of the skull. The reconstruction was then compared with a photograph of the man taken before he died, and it was shown to be successful with an accurate profile. The results of this experiment somewhat quietened Gerasimov's critics. Gerasimov further claimed that a definite resemblance was established in all of his 140 facial reconstruction cases. Unfortunately these studies are only vaguely recorded and the establishment of similarity appears to have been accepted from Gerasimov's subjective assessment. Some theorists (Tyrell *et al.*, 1997) have cast aspersions on the accuracy of Gerasimov's records and have suggested that his work was altered after the individual

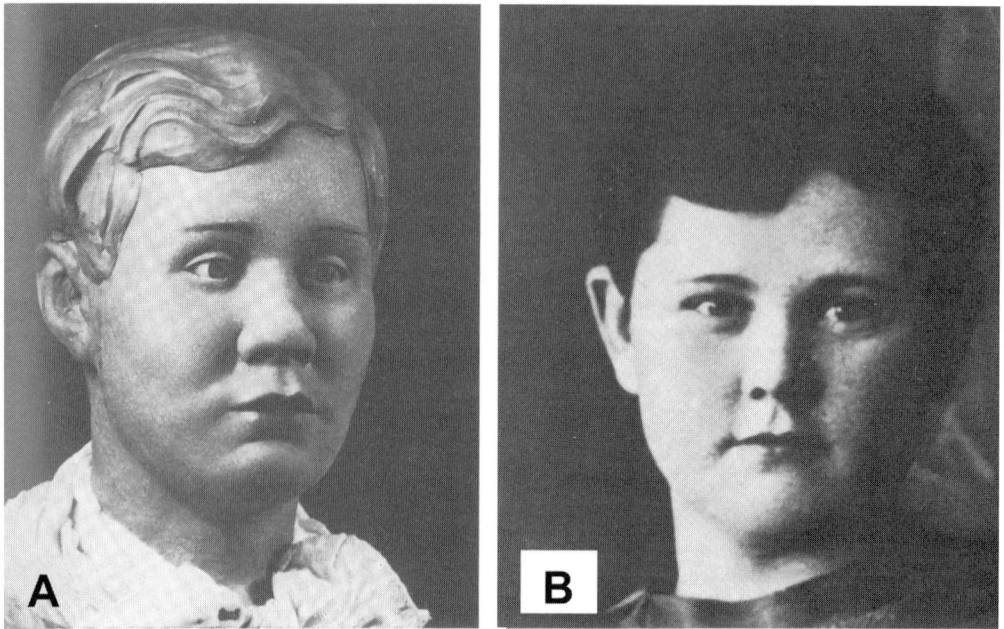

Fig. 7.3 The facial reconstruction of Valentina Kosova by Gerasimov, 1971. Reconstruction (A) and identified individual (B). Courtesy of the Laboratory of Anthropological Reconstruction, Moscow. From *The Face Finder* by M. Gerasimov, published by Hutchinson. Used by permission of The Random House Group Limited.

was identified. There were certainly fantastic results in his work. Gerasimov managed to sculpt the hairlines and hairstyles of many of his cases correctly, although the forensic remains did not show such detail. He claimed to have modelled hairstyles from verbal descriptions only when the police had a suspect for the victim. In one remarkable case he correctly reconstructed the jawline of a 22-year-old woman, Valentina Kosova, even when there was no mandible upon which to base the reconstruction (see Fig. 7.3). Gerasimov uses two particular cases to establish his accuracy: that of Valentina Kosova and the case of a 32-year-old woman, Nina Z. When studying the skull of Nina Z he noted atrophic changes to the '*cheek–forehead suture on the right side*', which he explained as a right-sided facial paresis. He concluded that this would have been very noticeable in the upper right eyelid, which would have drooped considerably. The resulting reconstruction was very similar to the identified woman,

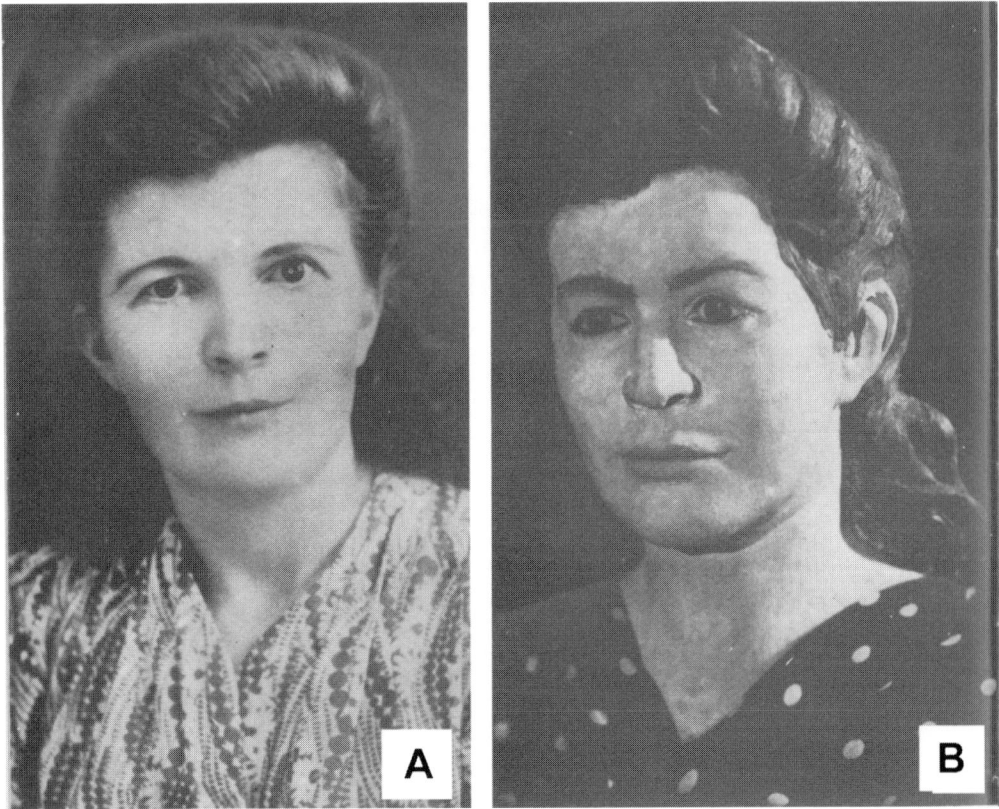

Fig. 7.4 Facial reconstruction of Nina Z by Gerasimov (1971). Reconstruction (B) and identified individual (A). Courtesy of the Laboratory of Anthropological Reconstruction, Moscow. From *The Face Finder* by M. Gerasimov, published by Hutchinson. Used by permission of The Random House Group Limited.

and the shape of the mouth and nose was well reconstructed (see Fig. 7.4). The eyes were almost identical with the right drooping lid. This reconstruction was photographed and shown to her husband, who admitted that it was a portrait of his wife. The reconstruction was accepted as accurate and tested with photographic super-imposition and odontology. The husband was charged and convicted of her murder. Gerasimov stated that '*attempts have been made to attribute to some personal talent of mine, and especially to some phenomenally developed sensitivity, the resemblance of reconstructions from skulls with photographs of their owners. All this is certainly very flattering for me but does not correspond to the facts.*' He stated that the success of his students following the technique was proof that it was the technique, not the reconstructor, that created a good resemblance.

In 1967 the North American artist, Betty Pat Gatliff, produced her first forensic facial reconstruction from the unidentified remains of a young man. After a great deal of media publicity the individual was recognised and identified as a young Native American (see Fig. 2.10). There has also been accuracy research into computerised facial reconstruction systems. Vanezis *et al.* (1989) compared the manual and the computer technique in a single-blind study. The results showed that both techniques could provide a useful tool for identification, although the manual technique produced a face that appeared more recognisable when compared with photographs of the individual. Rathburn (1984) studied the case of an unidentified 24-year-old black woman of medium build. Using the Rhine and Campbell (1980) Black American measurements, a facial reconstruction of the skull was carried out. Recognition came from a butcher who thought the reconstruction resembled one of his customer's daughters who had gone missing a year earlier. Comparison with a photograph of the individual showed a reasonable likeness and she was eventually identified by fingerprints. Haglund and Reay (1991) carried out an experiment to evaluate facial reconstruction techniques in identification of the Green River serial murder victims. A total of 24 reconstructions were produced after conventional methods of identification had failed. Nine different artists were used and information regarding the age, sex, racial group, height and date of death were provided. The artists used either 3-D or 2-D methods. Additional information regarding the hair colour and length was provided where known. The results showed that the interpretation of the same victim varied greatly and resemblance to the deceased showed considerable variation. None of the reconstructions elicited information that led to identification. However, some of the reconstructions were quite accurate and it was concluded that facial reconstruction can be useful for forensic identification. Haglund and Reay stated that '*although resemblance to the deceased is desired, this goal is rarely achieved . . . and unrealistic expectations among both the public and the investigators have been created'.*

Hundreds of international forensic cases have used facial reconstruction to produce identification of the individual through recognition. The Russian method, as practised by Gerasimov (1971), claimed a 100 per cent success rate. The American method, as practised by Gatliff (1984), claimed a 65 per cent success rate, and Helmer *et al.* (1989) a 50 per cent success rate. The British method, as pioneered by Neave (Wilkinson & Neave, 2001), claimed a 75 per cent success rate. However, in the majority of forensic cases, similarity of the reconstruction to the identified individual is established by a subjective visual assessment by the reconstructor alone. More quantitative methods of comparison have therefore been developed.

Quantitative studies

Some studies attempted to introduce quantitative comparison by using anthropometry. In 1946 the American anthropologist, Krogman, employed an artist, McCue, to restore the face of an individual. Krogman used the skull of a Black American man of 40 years, and McCue carried out the facial reconstruction using soft tissue data collected by Todd and Lindala (1928) from American and Malaysian Blacks, having been given the sex, age and race of the individual. Krogman, meanwhile, had measured the soft tissue from the head of the cadaver. The reconstruction (see Fig. 2.9) was considered to be readily recognisable and although there were errors at the bipalpebral breadth (13 mm underestimation) and bigonial points (10.5 mm overestimation), due to the inadequate anatomical knowledge of the sculptor, it was considered to be a useful technique. Research into a computerised facial reconstruction system was carried out by Gonzalez-Figueroa (1996) using 19 skulls of unidentified missing people. The computerised reconstructions were compared with 22 photographs of missing people by photogrammetry and computerised anthropometry. Samples of

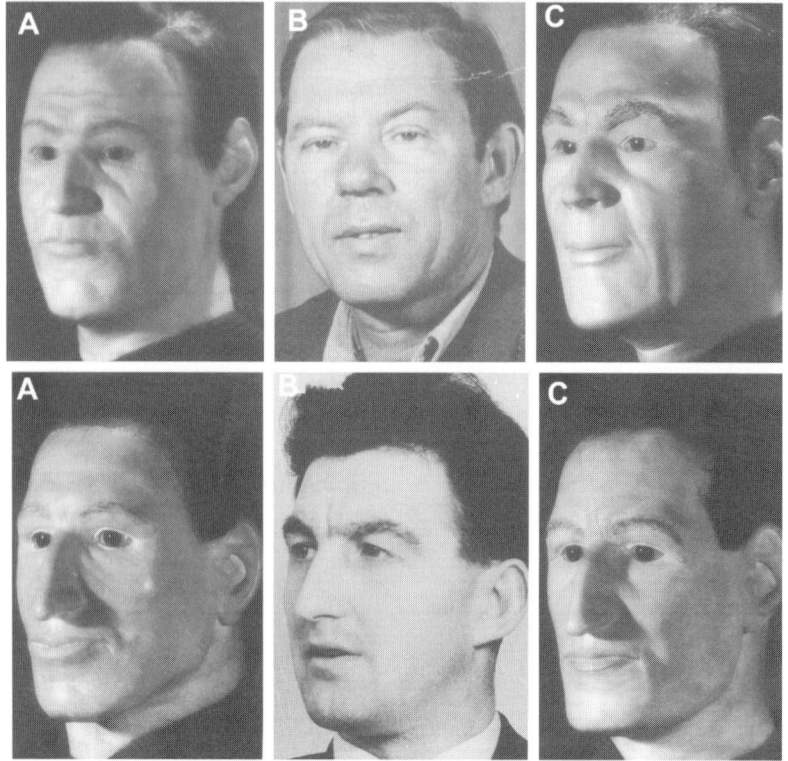

Fig. 7.5 Facial reconstruction reliability assessment by Richard Helmer (1993). Two cases (top and bottom rows) show two reconstructions (A and C) of each identified individual (B). From *Forensic Analysis of the Skull* by Iscan and Helmer. © 1993 Richard Helmer. Reprinted with permission from Wiley-Liss Inc., a subsidiary of John Wiley & Sons, Inc.

deoxyribonucleic acid (DNA) confirmed identification. This study showed that the facial proportion indices were significantly different in 48 per cent of the cases and not significantly different in 52 per cent of the cases. The researcher concluded that although these results were not particularly accurate, the technique showed some promise and feasibility.

Some studies attempted to assess the accuracy of the reconstructions scientifically, without losing the qualitative comparison of the face. A double-blind study was carried out by Helmer *et al.* (1993) (see Fig. 7.5). Helmer took two examiners who reconstructed 12 skulls of known age, sex and stature following a double-blind reconstruction plan based upon the morphology of the skull. He found that comparison of the reconstructions with each other showed 50 per cent of the cases with approximate resemblance and 33 per cent with close resemblance. Comparison of the reconstructions with photographs of the individuals showed 38 per cent close resemblance, 17 per cent approximate resemblance and 42 per cent slight resemblance. There was no resemblance in only one case. Helmer stated that artistic training of the reconstructor is not necessary, as it would eliminate the chance of negative results being influenced by artistic feeling.

There have been suggestions (Stephan, 2000) that the resemblance rating method of assessment for judging the accuracy of a facial reconstruction may be flawed. Stephan found that when subjects were asked to identify the appropriate individual approximated by a reconstruction from a face pool, and then to rate the resemblance of the reconstruction to the identified individual, the resemblance ratings did not differ between identifications that were true positive (where the volunteer correctly identified the individual) and false positive (where the volunteer incorrectly identified the individual). He concluded that resemblance ratings do not indicate the accuracy of a facial reconstruction. However, Stephan's results do not seem surprising since those individuals with false positive identifications presumably thought that the reconstruction looked most like the falsely identified individual, and it seems reasonable to assume that they would then rate that comparison in a similar way to the individuals with true positive identifications. More interesting would be the results of such a rating assessment when the subjects were asked to rate the reconstruction against the target individual, regardless of whether or not they chose the target individual in the face pool study. Wilkinson and Whittaker (2002) studied the reliability of resemblance ratings for the assessment of facial reconstruction. Wilkinson produced five reconstructions, and then a photographic face pool of ten individuals of similar sex, ethnic origin and age was set up, including the five targets. Fifty volunteers chose the face from the face pool that most resembled each reconstruction, and then the same volunteers were asked to

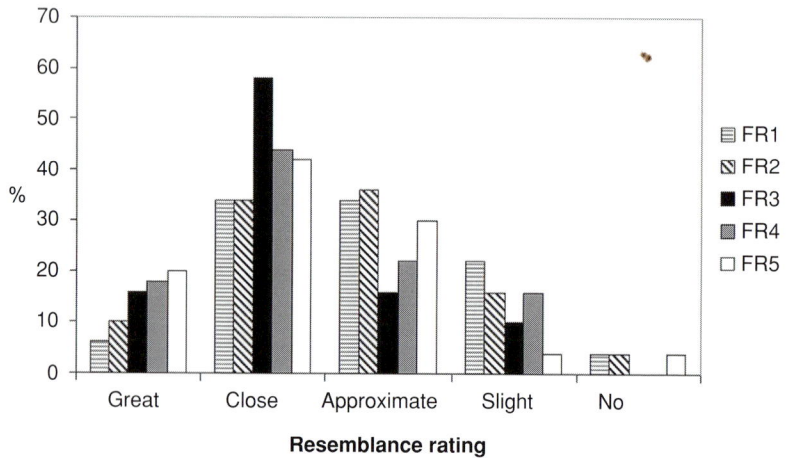

Fig. 7.6 The resemblance ratings for facial reconstructions (FR) when compared to identified individuals (Wilkinson & Whittaker, 2002).

rate the reconstructions as a resemblance of the target individuals using a five-point scale, from no resemblance to great resemblance. Included in this resemblance study was a 'foil' comparison where the individual was not the target, but an unrelated face from the face pool. The overall likeness ratings for the reconstructions and the target individuals were 14% great, 42% close, 28% approximate, 14% slight and 2% no resemblance (see Fig. 7.6). At all points the most frequently chosen rating was close resemblance. The foil comparison was rated as 48% slight and 40% no resemblance. With respect to the body fat, the nose, the mouth, the chin and the overall likeness, the foil showed significantly worse resemblance ratings than all the other comparisons (see Fig. 7.7). In contrast to the deductions of Stephan, Wilkinson and Whittaker concluded that resemblance ratings are an accurate method of assessment for facial reconstruction. All five reconstructions were rated as close overall resemblances to the identified individuals. The age and the mouth resemblances were the lowest ratings and the body constitution, eyes and nose resemblances were the highest ratings. One of the problems with resemblance-rating research is that it is still a subjective assessment of similarity, and different observers have very different responses to the same reconstruction.

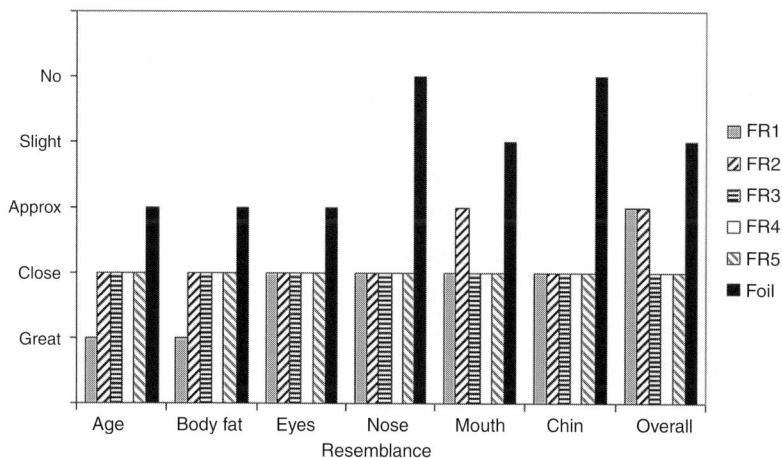

Fig. 7.7 Comparison of the resemblance results for the facial reconstruction identifications (FR) with the foil identification (Wilkinson & Whittaker, 2002).

In addition, it is thought that observers tend to look for errors in the reconstruction when comparing it to the identified individual rather than similarities. Therefore, the method of face pool identification was developed to assess the accuracy of a facial reconstruction.

Face pool identification involves the collection of a number of photographs of individuals similar to the identified individual (target), in order to create a face pool from which volunteers choose the face that most resembles the reconstruction. The similarity between the faces in the face pool may involve sex, age, ethnic origin, face shape or face type and often the images in the face pool are similar in view, angle and composition. These studies produce results that show the percentage of correct identifications (hits) and these can be compared to the number of correct hits that would be produced by chance. However, these studies do not wholly reflect a forensic scenario since the assessments are of unfamiliar faces. In forensic investigations the facial reconstruction will usually be recognised by a family member or close friend of the individual, and psychology research suggests that we recognise unfamiliar faces in a different way to familiar faces (Ellis *et al.*, 1979; Shepherd, 1981; Bruce & Young, 1998). Despite this, the face

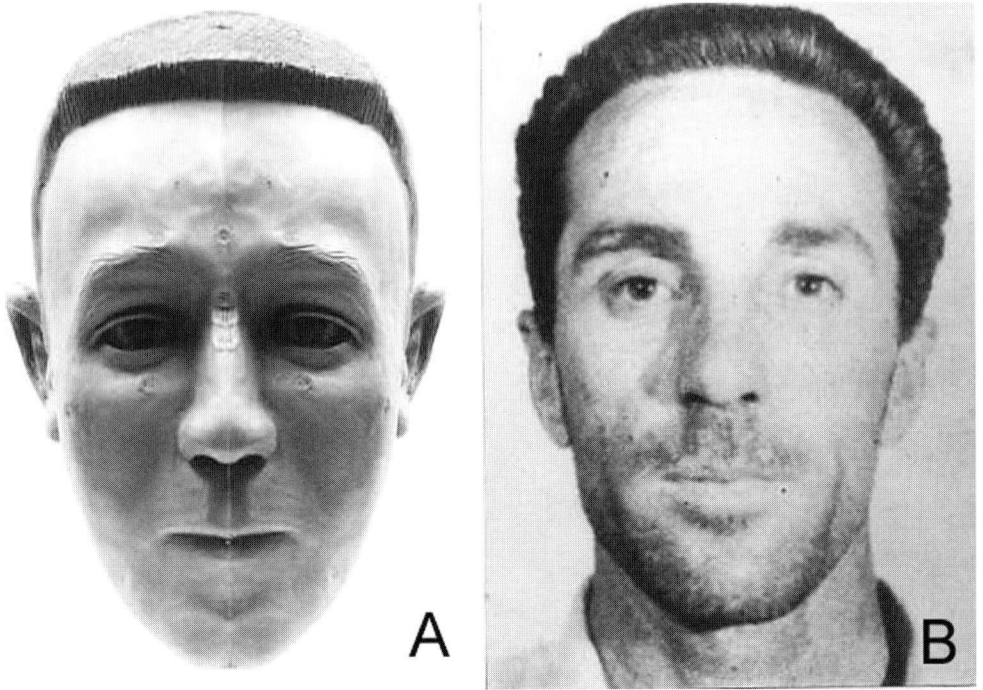

Fig. 7.8 Facial reconstruction (A) of 36-year-old man by Gatliff, and identified individual (B). Courtesy of Karen Taylor (2001).

pool identification study appears to be the method of assessment closest to a forensic investigation. Snow *et al.* (1970) carried out a face pool identification appraisal of the American method, using Gatliff as the reconstruction artist, producing two reconstructions of a white male, aged 36 and a white female, aged 67. Posters were constructed using the clay reconstructions and seven photographic portraits of individuals of similar sex and age, and volunteers were asked to select the individual who most resembled each reconstruction (see Fig. 7.8). The female scored approximately 25% correct selections (11% above chance) and the male 68% (54% above chance). This result was somewhat confusing and, in addition, only half the face had been reconstructed so that the reconstruction was created by combining the reconstructed half with its mirror image. This process produces an unnaturally symmetrical face and an unnatural lighting effect. The female result was poor, but all the photographs of the women were taken at the age of 42 and the reconstruction was of a woman 25 years senior. However, the male result suggested a significant potential for this technique. Gatliff and Snow (1979) therefore expressed guarded optimism that a reconstruction may produce a face that bears a fundamental resemblance to the unknown individual. Van Rensburg (1993) used 11 judges to attempt to identify 15 reconstructions by face pool identification. This method unusually employed death masks for comparison with the reconstructions. A total of 40% (33% above chance) of the reconstructions were correctly identified by comparison of one death mask with one reconstruction at a time; 17% (10% above chance) were correctly identified from comparison of one reconstruction with all the death masks; and 19% (12% above chance) were correctly identified from comparison of all the reconstructions with all the death masks together. The average correct identification was 19% above chance. Stephan and Henneberg (2001) carried out a study that used 37 assessors who attempted to identify 16 reconstructions by face pool assessment. Four skulls were reconstructed using four different methods: 3-D American, 3-D British, 2-D American drawn, and computer-assisted methods. The face pool consisted of ten individuals. They found that only one of 16 facial reconstructions was identified at a significant hit rate above chance, and recorded an overall mean hit rate of only 6%, which was 19% below chance. These results suggest that facial reconstruction does not produce a good likeness to an individual and that it would be detrimental to any forensic identification case. However, it must be noted that the reconstructor had very little experience. Wilkinson and Whittaker (2002) carried out a detailed accuracy study of forensic facial reconstruction using five female juvenile skulls between the ages of eight and 18 years. The reconstructions were produced employing tissue depth data from White British children (Wilkinson, 2002). A photographic face pool of ten juvenile white females was set up, including the five target individuals. Fifty volunteers chose the face from the face pool that most resembled each reconstruction. The five reconstructions were all correctly identified as the most frequently chosen face from the

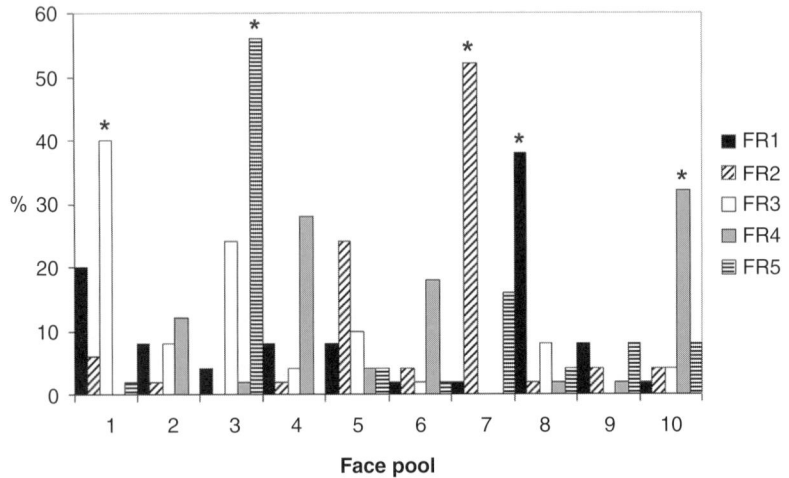

Fig. 7.9 Facial reconstruction accuracy test using face pool identification. Each reconstruction (FR) was correctly identified as the most frequently chosen face (*) from the face pool (1–10). Wilkinson & Whittaker (2002).

face pool, and the mean hit rate was 44%, with all hit rates well above chance (10%). The individual hit rates for each reconstruction were 38%, 52%, 40%, 32% and 56% (see Fig. 7.9). Wilkinson and Whittaker concluded that it was possible to create a good likeness of an individual following the Manchester method of facial reconstruction.

The results of these face pool identification studies are put into perspective by the results of research by Bruce *et al.* (1991; 1999). Bruce and her colleagues studied the recognition of 3-D facial surfaces using laser-scanned images as the head model. A face pool of photographs of four males of similar age, one of whom was the individual who had been scanned, was set up and the subject was asked to pick out the individual whom the scan represented. Despite the fact that the scan represented a face of exactly the same proportions, shape and size as one of the individuals, the overall hit rate was only 51% (26% above chance). In addition, they investigated matching of unfamiliar target faces from high-quality video stills against arrays of ten photographs of similar aged males, one of whom was the same individual as seen in the video still.

The target faces were shown either in frontal, three-quarter or profile views, whilst the array faces were all shown in frontal view (see Fig. 1.18). The overall hit rates were only 70% (frontal – neutral face), 64% (frontal – smiling) and 61% (three-quarter face), despite the fact that the target photograph was taken the same day as the array photograph of the same individual. This result surprised even the researchers. Clearly we have difficulties with recognising faces, even when a direct comparison is carried out; the results of the face pool assessments for facial reconstruction can, therefore, be considered very positive with regard to the possible accuracy produced. There are, however, problems with all the accuracy assessments that use photographic images of the identified individual. Often the images available in forensic scenarios are poor quality and/or taken from awkward angles or views, and there may be an age difference between when the photograph was taken and the age at death. People often smile when having a photograph taken and this can make an assessment of the mouth and cheeks difficult, and lighting in the photograph may obscure or overemphasise details of the face.

Neave (Prag & Neave, 1997) tried to circumvent the problems associated with photographic comparison by reconstructing the face of someone who was alive. With the development of medical imaging it is now possible to CT scan a person and manufacture a copy of the individual's skull using a process known as stereolithography (Hjalgrim et al., 1995). CT scans irradiate the subject and there is some risk involved, since many scans are necessary to produce a good copy of the skull. This type of procedure requires volunteers in the later years of their life and is, therefore, inappropriate for many studies in this area. Therefore, opportunity to carry out such research is rare. As part of a challenge set up by the Dutch National Association for Oral and Maxillofacial Surgery, Neave was provided with a styrene foam copy of a skull made from CT scans of a volunteer's head using a computer-controlled laser-cutting technique (see Fig. 5.5). The only information supplied to Neave was that the individual was '*Dutch, male, retired and with little*

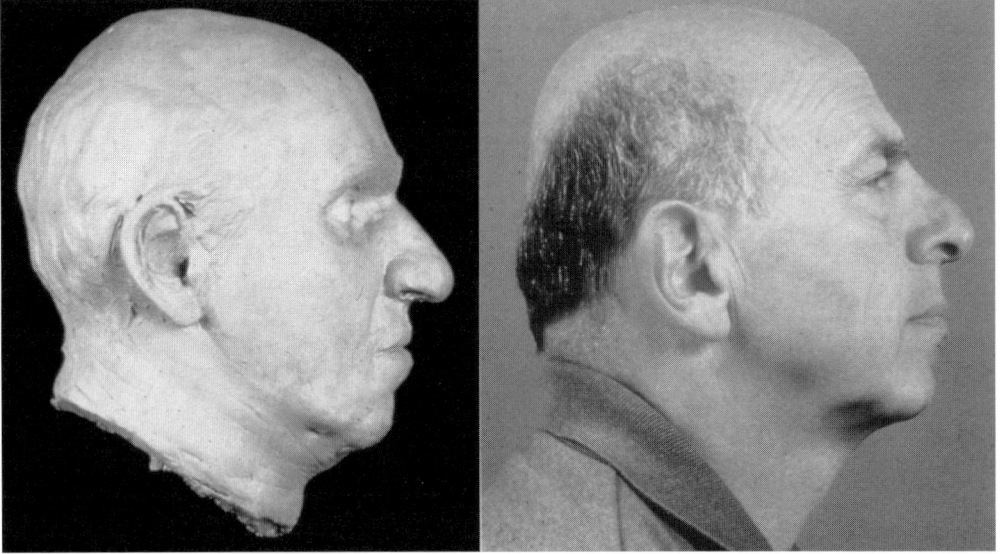

Fig. 7.10 Comparison of facial reconstruction (left) and actual individual (right) by Neave (1997). Modified from Prag and Neave (1997).

hair'. The reconstruction was sufficiently similar to the individual for Neave to recognise the volunteer, Professor Peter Egyedi, in a room full of people (see Fig. 7.10). Neave stated, '*Of course the face is not exactly the same as Professor Egyedi's – one never claims that the reconstructions are portraits, after all – but it was very similar; similar enough for me to understand what happens when someone recognises the reconstructed face of a person whom he or she knows, even though in my case it was the other way round.*'

In conclusion, most of the results of facial reconstruction accuracy studies suggest that it is possible to recreate the face of an individual from the skull with sufficient enough accuracy that someone who knew them well, such as a family member or close friend, would be able to recognise them. However, these results also suggest that the reconstruction practitioner must be experienced and have a thorough understanding of facial anatomy, anthropology, physiology and related pathology. In addition, the reconstructor must have practical skills in sculpture and a specialised knowledge regarding the relationship between the soft and hard tissues of the face. It is clear that it is possible to train someone in the technique of facial reconstruction very quickly, but that accuracy of characteristic facial reconstruction only develops with extensive study, experience and rigorous scientific method.

8 Juvenile facial reconstruction

The reconstruction of children's faces has always been considered more problematic than adult facial reconstruction. There are many difficulties associated with juvenile remains, including more inaccurate sex and racial origin determination, the more emotive and sensitive nature of an investigation into the death of an unknown child, and the less defined skeletal details associated with underdeveloped skulls. Historically this field has not been separated from adult facial reconstruction, although the differences between adult and juvenile skulls are enormous. This may be a reflection of the small numbers of juvenile reconstructions produced relative to adult reconstructions, and it may be due to the paucity of juvenile research. Until 1963 there were no published juvenile facial tissue depth measurements, and anthropometrical studies have concentrated on facial growth patterns rather than facial standards. However, there has been a great deal of anthropological research into age and sex determination, and the growth of juvenile skulls.

Juvenile age determination

With the remains of preadolescent children there are three age estimates that can be used: the numbers of emerged teeth, calcification of the permanent mandibular first molar, and the schedule of calcification of the dentition as a whole. A consistent and predictable sequence has been determined for the formation, eruption and loss of primary dentition, and its replacement with secondary dentition. The rates have been calculated for children from many populations (Iscan & Helmer, 1993). All the deciduous dentition and the first permanent molars have begun to mineralise at birth. The deciduous dentition has emerged and rooted by the age of approximately three years, and during the first year the permanent first molar and anterior teeth begin to form. The premolars and second molars mineralise between the ages of two and four years, and the third molars begin formation between the ages of six and 12 years. The emergence of all the permanent teeth, except the third molars, takes place in two stages between six and eight years, and then again between ten and 12 years. The first permanent molar emerges behind the second deciduous molar, whilst the deciduous incisors are shed and replaced by the permanent teeth. The pattern of emergence is usually mandibular central, maxillary central, mandibular lateral and maxillary lateral incisors. The second stage includes the shedding of the canines and premolars to be replaced with the permanent canines, premolars and second molars. The third molars appear in adolescence or early adulthood and are more variable in size, shape and emergence. Schedules of eruption can be found in Lysell *et al.* (1962); Meredith (1973); Ubelaker (1978); Liversidge *et al.* (1998) (see Fig. 8.1). Accurate age determination is possible up to the age of 12 years.

The growth of the skull throughout a lifetime leads to enormous changes in our facial appearance from birth to adulthood (see Fig. 8.2). Y'Edynak and Iscan (1993) stated that the human face goes through a series of changes so drastic during the first years of life that an infant might be unrecognisable if not seen for even a few months. Other changes such as skin elasticity, fat distribution, and texture and pigmentation of the skin also lead to age-related changes in the juvenile face. The examination of lifetime facial growth has been constantly researched from as early as the Italian Renaissance. A brief overview of the facial growth pattern from birth to death may facilitate understanding of the problems associated with facial reconstruction in relation to age.

The head of an adult constitutes approximately one-eighth of its total body length, whereas that of an infant constitutes a quarter. The youthful face appears to be more brachycephalic, since it is relatively short and wide. This is due to the large brain, which is

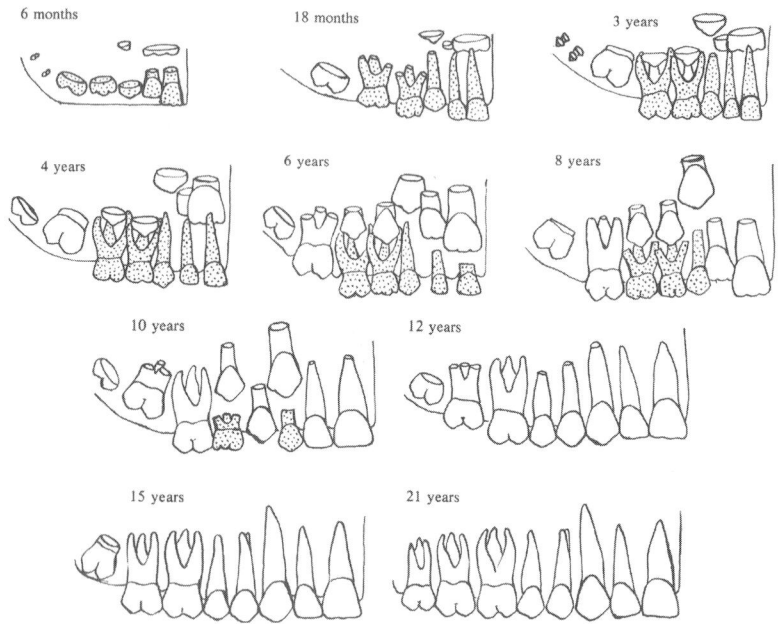

Fig. 8.1 Average developmental stages of human dentition from six months to 21 years of age. Stippled teeth represent the deciduous dentition. Modified from *Digging up Bones* by Brothwell (1981). Used by permission of the publisher, Cornell University Press.

precocious in development relative to the face. Enlow (1982) explained that the infant has a vertically short face because of the diminutive nasal and oral regions that match the body and lung size, and because masticatory development is in a transitory state. The dental and nasal regions are late developing and, therefore, the ramus of the mandible is short, leading to the vertically short appearance. Nasal enlargement keeps pace with the growing body and lung size, and mandible configuration keeps pace with dental development. Regardless of sex, the child's face has a short, pug-like nose with low nasal bridge and concave profile. The infant nose gives no indication of the adult nasal shape and size. The nose of a child is vertically shallow, due to the diminutive nasal chambers and small lung size. The superior and inferior orbital margins of a child are in an approximately vertical line, or inclined posteriorly,

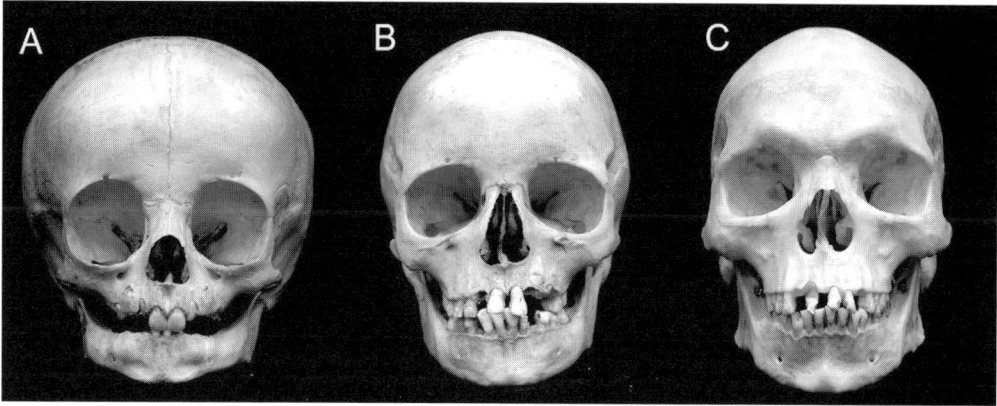

Fig. 8.2 Age-related changes to the skull. A = nine months old, B = 9–10 years old, C = adult male.

and frontal sinus development will lead to supraorbital protrusion in the adult. The juvenile forehead is bulbous and upright,
the cheekbones prominent, the face flat and the eyes wide-set and
bulging. The baby face appears to have large eyes, a dainty jaw,
puffy cheeks, low-placed ears, a small mouth and a high forehead.
A child's cheeks have larger buccal fat pads than the adult, which
give the child a smooth-skinned, round-cheeked appearance. These
features gradually undergo marked changes as the face develops
and grows (see Fig. 8.3). The chin develops, the jaw catches up in
size, the eyes appear less wide-set, the face enlarges inferiorly more
rapidly and the proportionate size of the forehead decreases. The
sinuses enlarge, the jaw becomes squarer, and the gonial area becomes more lateral. The eyes appear larger in the child, and they
look progressively smaller, relative to the face, as the child grows
in age. Eyeball size is precocious to the facial development, but the
eyes are actually similarly spaced in adults and children. The larger
nose, higher nasal bridge, longer face and wider cheekbones will all
cause adult eyes to appear closer together. The ears of the child
appear low and, as the face elongates, they appear to rise. In fact the
ears move downward during development, but the face enlarges
even further downwards, so that the relative position of the ears

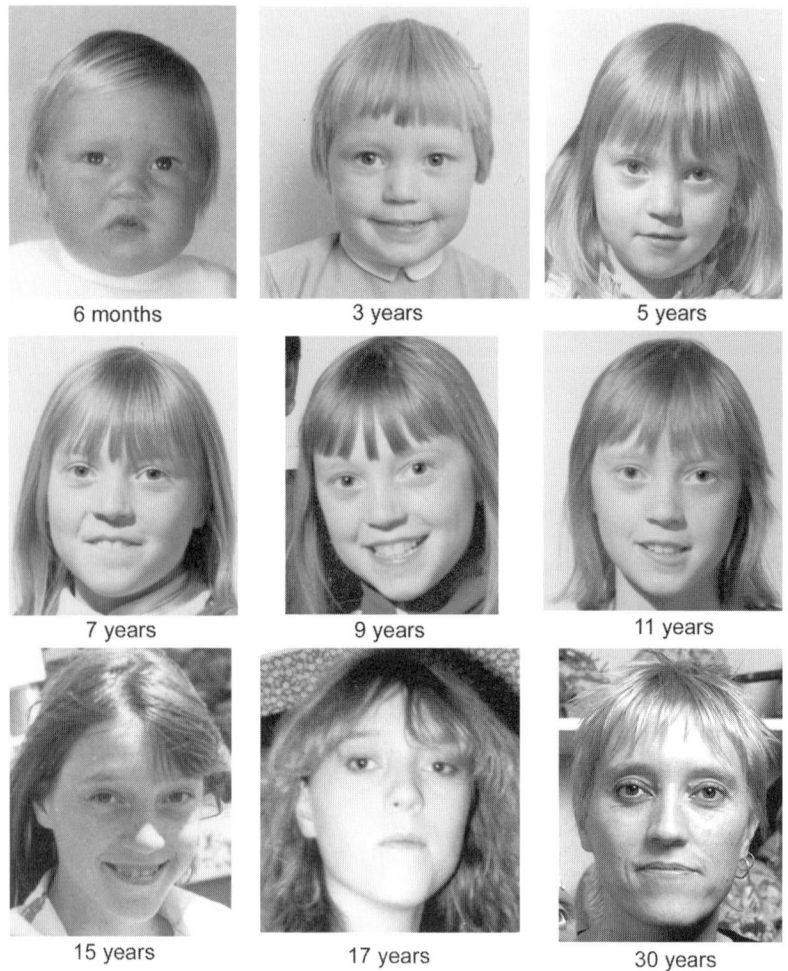

Fig. 8.3 Age-related changes to the juvenile face. The same individual between the ages of six months and 30 years.

rises. The child's forehead is much more bulbous and becomes gradually more sloping with age. The nasal bridge of the child is low and gives the impression of wide-set eyes. As age increases the nasal bridge becomes higher and the cheekbones wider, so that the eyes appear more narrow-set. In addition, the mandible of the child is small and underdeveloped relative to the face in general, and the chin is barely formed in an infant. The anterior cranial fossae are

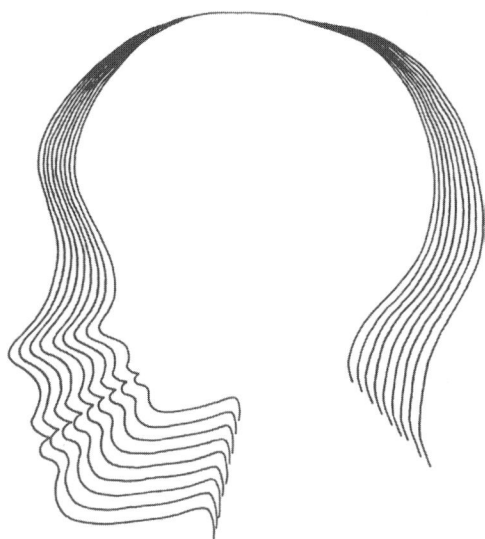

Fig. 8.4 Cardioidal-strain transformation profiles of craniometric growth of an infant (innermost profile) to an adult (outermost profile). Modified from Todd et al. (1980).

developmentally precocious, so the nasomaxillary area is in a more protrusive position than the mandible. The young child's chin appears pointed, and becomes squarer with age. The emergence of teeth has a large effect upon the face, and, as each tooth emerges in the infant, or is replaced in the young child, the face alters in form slightly. The development of the chin, eruption of the teeth, enlargement of the ramus of the mandible, expansion of the masticatory muscles and the flaring of the gonial angles lead to the whole lower face taking on a more U-shaped appearance. The small mastoid processes of the child's skull may develop into quite large protuberances in the adult. The adult face is longer, more sloping, the face is proportionally bigger relative to the head, and the general appearance is more dolichocephalic. These facial changes are illustrated in Fig. 8.4.

Farkas and Posnick (1992) studied facial growth in American children, and found that the face reaches maturity by about ten to 13 years in girls, and 12 to 15 years in boys. El-Nofely (1972) studied Nubian children, and indicated that between the ages of six and

12 years cephalic length increases continually in both sexes. He also found that cephalic breadth reaches a peak velocity at nine to ten years in boys, whilst girls are characterised by slow periods at eight to nine years and ten to 11 years. Farkas and Posnick (1992) stated that full maturity in head length, breadth and circumference occurred at age ten years in girls, and 14 years in boys. The adolescent growth spurt leads to a faster mandible growth in boys than girls, producing larger, more projecting jaws in men than women. A similar growth spurt in the nose occurs in boys, creating larger, more prominent noses in men than women.

Anthropometry studies (Bjork, 1947; Farkas & Hreczko, 1994; Tanner, 1952; Feik & Glover, 1998) also demonstrate enormous growth changes in the face from infancy to adulthood. At one year of age the head and orbits are 78–89% of their adult size, but the nasal and mouth dimensions are only 50–73% of their adult size. In both sexes to achieve their adult sizes, mandible height has to increase by a mean of 50%, mouth width by 52%, and nasal tip protrusion by 96%, of their initial sizes at one year of age. Head length, ear width and intercanthal distance show much lower growth increases (7–22%) in the same period. Facial features showing small early development, such as nasal tip protrusion (51%), have rapid growth during the first seven years of infancy in both sexes, whereas facial features with high early development, such as head width (84%) and interocular distance (86%), have steady growth with below-average increases. These studies also found that forehead inclination decreased (mean 13.7°), upper lip incline decreased (mean 11.9°) and nasal tip inclination decreased (mean 9.8°) with an age increase. The columella inclination decreased (mean 17.8°), lower lip inclination increased (mean 12.6°), and nasofrontal angle increased (mean 7.9°) with an age increase. The facial features that reached adult dimensions earliest were the ear width (female at six years, male at seven years) and head length (female at ten years, male at 14 years). The features with high (80–90%) developmental levels (intergonial and intercanthal distances, and head width) reached mature sizes between ages five and 15 years. Facial features with lower (70–80%) developmental levels (zygomatic and interalar distances) matured between ages 11 and 15 years, and those with the lowest (55–70%) developmental levels (mouth width, lower face height and nasal length) matured between ages 12 and 15 years. Other 3-D studies of facial growth (Nute & Moss, 2000) showed that face height increased by approximately 4 mm a year, lower face width increased by approximately 1–3 mm a year, and the intergonial width increased by approximately 3–5 mm a year, between the ages of five and ten years. They also found that the chin prominence varied little until after the age of eight years, and nasal prominence increased more steeply after the age of eight years.

Juvenile sex determination

The determination of sex from the cranial remains of children is notoriously difficult and uncertain, as is the sex determination from juvenile skeletal remains in general. However, Hunt and Gleiser (1955) studied radiographs of the same children at ages two, five and eight years, and correctly sexed 73–81 per cent of the children by comparing the schedules of eruption, which were shown to be different between males and females. As the female moves from puberty to adulthood her skull retains much of the gracility and smoothness characteristic of a prepubertal skull, whereas the male becomes less gracile, larger and rougher. Some researchers (Krogman & Iscan, 1986) claim that between the ages of two and eight years there is 60–80 per cent accuracy in sex determination. But others (Stewart, 1948) suggest that juvenile sex determination is no better than a guess, at little more than 50 per cent. Some researchers (Acsadi & Nemeskeri, 1970; Schutowski, 1993; Molleson et al., 1998) demonstrated sexual dimorphism in the mandibles and orbits of juvenile skulls. It was found that the orbits were more rounded with sharp margins in females, and more angular with rounded margins in males. In addition, the female orbits were larger in relation to face size. The male mandible showed a more everted angle and squarer chin, with more defined muscle attachments. Molleson *et al.*, (1998) studied juvenile skulls from the medieval Spitalfields site in London, and correctly determined sex in 78 per cent, using orbital and mandibular morphological assessment. Some research (Garn *et al.*, 1964; Black, 1978) suggests that the dimensions of the tooth crown tend to be larger in males than females, especially the canines, and that this can be demonstrated in juvenile remains. Mays and Cox (2000) suggested that odontometry of the canine teeth can provide a potential sexing method from middle-to-late childhood. They stated that after the age of six years the crowns are complete for measurement purposes, and claimed that archaeological examples of adult sexing from dental measurements reported a success rate of 75–95 per cent (Ditch & Rose, 1972; Mays, 1996). However, it must be noted that nutrition and health may affect dental crown size, and the success rate of the sexing of juvenile archaeological samples by this method cannot be assessed due to the unavailability of independent sex indicators. The assessment of DNA may be another potential method of sex determination.

Any differences between the skulls of boys and girls will be reflected in their faces, and there has been a great deal of research into these differences. Anthropometry studies (Riola *et al.*, 1974; Farkas, 1981; Van der Beek *et al.*, 1991; Bishara, 1995) demonstrated sex differences in facial height varying from 1 mm to 6 mm, with boys' faces being greater in height than girls' faces. Zygomatic width and intergonial width measurements were shown (Riola *et al.*, 1974; Farkas, 1981) to be greater in boys than girls by 1 mm and 3 mm respectively. Some studies (Farkas, 1981; Burke & Hughes-Lawson, 1989) suggested that sex differences in nasal dimensions are small (maximum 1 mm) until nine years of age, after which the male nose grows at a greater rate. Three-dimensional studies (Nute & Moss, 2000) also demonstrated some sex differences in the faces of children, aged between five and ten years. A greater face height (7–9 mm) was seen in boys than girls, but a similar midface width. The mandibular width was greater (3–5 mm) for boys than girls in most age groups, and the intergonial width was greater (5–7 mm) for boys than girls at all ages, except the six-year-olds.

Boys had more prominent (1–3 mm) chins than girls at most age groups, but interorbital distances were similar for boys and girls. The nasal dimensions were similar in the five-to-seven-year-olds, but the ten-year-old boys showed wider interalar distances, height and prominence (1–3 mm) than girls.

As mentioned in Chapter 3, we are very good at determining the sex of an individual from the face. However, this only applies to adult faces. We are spectacularly bad at determining the sex of juvenile faces. The faces of prepubertal boys and girls are essentially comparable. Most people have, at least once in their life, mistaken a boy for a girl, or vice versa. The faces of babies illustrate the extreme of this lack of sexual dimorphism, and baby boys and girls are indistinguishable. The sexual characteristics associated with facial appearance develop during puberty. Enlow (1996) stated that female facial development slows after the age of 13 years, but the male face begins to manifest the sexually dimorphic facial features. A similar sex determination study, as previously described for adult faces (Bruce *et al.*, 1993), was carried out using juvenile faces, and the correct sex was indicated in only 52 per cent of the faces. This is little better than guesswork, and is a reflection of the predominance of female facial characteristics in both male and female children (see Fig. 8.5). The younger the child, the more difficult it is to determine the sex. Very young children tend to have similar features (upturned noses, chubby cheeks, brachycephalic head shapes etc.).

Juvenile racial origin determination

Racial origin determination from juvenile remains is so fraught with difficulty that the majority of forensic and physical anthropologists appear to steer clear of even mentioning this subject. This difficulty has arisen because the allometric changes in the craniofacial skeleton during postnatal growth is greater than the extent of the interpopulation morphology variation. Some researchers (Viðarsdóttir & O'Higgins, 2001; Viðarsdóttir *et al.*, 2002)

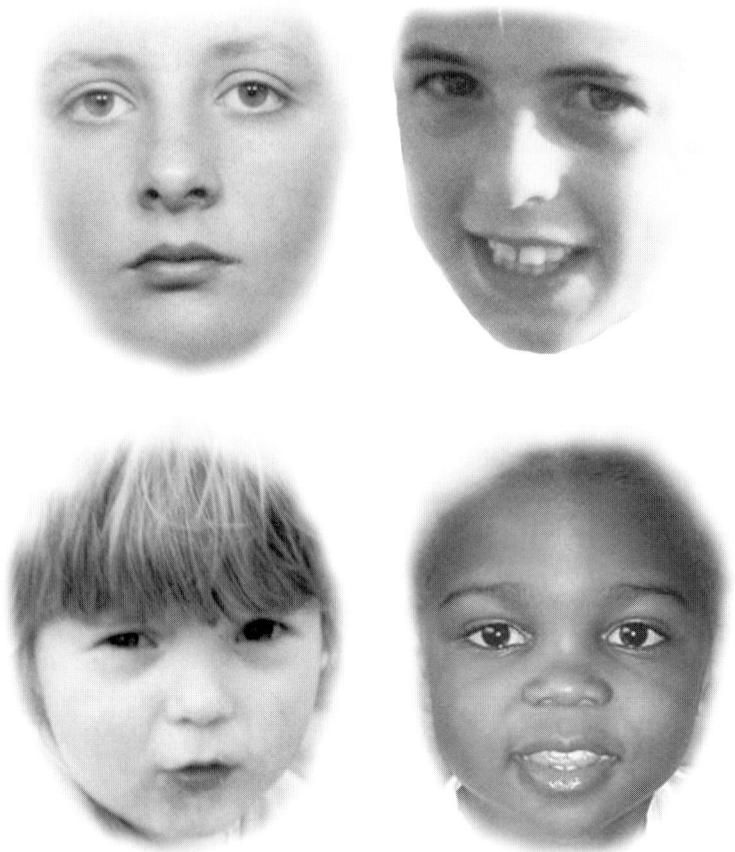

Fig. 8.5 Which of these children is male and female? On the top row are ten-year-olds and on the bottom row are two-year-olds. In both cases the girl is on the left and the boy on the right.

have attempted to overcome these problems with the use of advanced analytical techniques. The geometric 3-D technique examined morphological separation between ten modern ethnic groups with known morphological differences in adult facial form, including Africans, Europeans, North Americans, Melanesians, Australians, Alaskan Inupiaq Eskimos, Aleutians and Polynesians. They found that 71 per cent of the juveniles were correctly assigned to the correct population. Juvenile skulls tend to show some of their racial origin group characteristics, but these characteristics may be indistinct or ambiguous. Dental research (Bishara, 1995) demonstrated that most people (77 per cent) can be categorised as having the same basic face type (in relation to face length) at five and 25 years of age, suggesting that there is a strong tendency to maintain the overall facial pattern. These results suggest that general cranial parameters may be consistent in juveniles and adults and, therefore, we would expect to demonstrate racial group characteristics in skull length and width, sagittal contour, face length and width, and the degree of prognathism. The finer facial feature details may be more difficult to categorise. Nasal guttering may be present in juvenile Negroid skulls, but is often exhibited in some juvenile Caucasoid and Mongoloid skulls. Since infant noses tend to be short, upturned and snub-shaped, with nostril placement superior to the inferior nasal aperture border, they will also tend to exhibit ambiguous nasal guttering, regardless of racial origin group. As the nasal shape develops and the nasal projection, length and width increase, the nasal aperture shape will mature and become defined. Assessments of the degree of prognathism may also be difficult as most infant faces have prominent foreheads and dainty jaws and chins, which may detract from any inbuilt prognathism. While racial origin group determination may be difficult in early-to-middle childhood, adolescent skulls do tend to exhibit the majority of ancestral characteristics and can usually be categorised with a similar success rate (80 per cent) to adults.

Juvenile tissue depth measurement

Until very recently there were no useful data available related to juvenile facial tissue depths, for use in facial reconstruction. There were some radiographic studies (Altemus, 1963; Heglar & Parks, 1980; Dumont, 1986) produced, mainly for use in orthodontic research, but the traditional needle-puncture, MRI scan, CT scan and ultrasonic measurement techniques had been used to measure only adults. The early radiographic studies used lateral cephalographs, and were concentrated upon North American children (see Table 8.1). In 1963, Altemus carried out the first juvenile facial

Table 8.1 Radiographic facial tissue measurements (mm) from North American children.

Facial points	Altemus (1963) radiographic								Heglar and Parks (1980) radiographic						Dumont (1986) radiographic			
	White North Americans 13–16 years				Black North Americans 12–16 years				White North Americans 8–18 years						White Americans (93 males, 101 females)			
	Male (11)		Female (26)		Male (25)		Female (25)		Male (21)			Female (28)			9–11 years		12–15 years	
	Mean	SD	Mean	SD	Mean	SD	Mean	SD	Mean	SD	Range	Mean	SD	Range	Male Mean	Female Mean	Male Mean	Female Mean
Forehead									4.9	2.60	3–7	4.9	2.80	3.5–7				
Glabella	7.0	1.11	6.6	0.82	6.6	1.17	6.5	1.22	6.5	3.20	4.5–8.3	6.1	3.70	4.4–7.3	6.7	6.7	6.7	6.7
Nasion									6.2	3.30	4–8.5	5.6	3.60	3.5–7.7	7.3	6.8	7.3	6.8
Midnasal bone									4.9	2.80	3.8–7	4.3	3.70	2.2–6.7	4.6	4.0	4.6[a]	4.0
Rhinion									3.1	2.00	2–4.2	2.6	2.00	1.5–4.5	3.2	2.7	3.2[a]	2.7
Midphiltrum	16.2	1.61	14.7	1.88	17.8	2.49	16.5	2.11	12.9	8.90	8.7–20.5	12.7	5.60	10.2–16.2				
Upper lip	15.5[a]	1.88	12.1	1.83	16.4	3.04	14.2	2.43	14.5	6.20	11.6–17.5	12.5	4.00	10.2–15.4	13.7	13.4	14.9[a]	12.6
Lower lip	16.1[a]	1.54	13.4	1.29	17.1[a]	2.57	14.5	1.97										
Labiomental	12.9	2.20	11.6	1.31	15.1	3.11	13.0	2.01	11.0	5.50	8–14.5	11.7	4.70	9.6–14.8	12.1	11.7	12.8[a]	12.1
Mental tubercle									14.3	6.10	10.2–18.2	13.5	5.70	10–17.5	8.6	8.6	8.6	8.6
Gnathion															7.4	7.4	7.4	7.4

[a] Significantly thicker by sex.

Note: SD = Standard deviation

tissue study on White and Black North Americans between the ages of 12 and 16 years. He found that there was a wide variation in tissue depth at all points for both Black and White children, but the data were confined to midline measurements. Heglar and Parks (1980) continued this research by studying White American children, aged between eight and 18 years. Their results did not attempt to distinguish between age groups, and were, again, limited to midline data. Other radiographic studies followed (Dumont, 1986; Garlie & Saunders, 1999; Smith & Buschang, 2001; Williamson *et al.*, 2002) using North American and Canadian children (see Tables 8.1, 8.2 and 8.3). All these studies suffered from similar limitations in that the data were only midline, the subjects were lying down when the measurements were taken, and a wide variety of different tissue measurements were taken, not all relevant to facial reconstruction. Since the cheek and jawline areas of the face are demonstrably different between adults and children, further studies that measured these regions were necessary. It was not until 1985 that more complete sets of tissue depth data were produced for juvenile faces. Hodson and her colleagues (1985) studied American White children, between the ages of four and 15 years, using diagnostic ultrasound (see Table 8.4). However, this study also placed the subjects in a supine position during data collection. So it was not until 2000, when Manhein and her colleagues studied North American children, using ultrasonic imaging, that a complete and useful tissue reference for children's facial reconstruction was created (see Tables 8.5–7). They studied 515 White, Black and Hispanic children, between the ages of three and 18 years. A recent addition to this tissue data collection was produced, at the University of Manchester (Wilkinson, 2002), with a study of 200 White British children (see Table 8.8).

Differences between juvenile and adult tissue depths

When the North American study of juveniles and adults (Manhein *et al.*, 2000) was assessed, it was demonstrated that White children had thinner tissue in general than White adults, except at the

Table 8.2 Facial tissue measurements (mm) from White North American children aged 8–18 years.

Facial points	8–10 years Male (83) Mean	8–10 years Female (80) Mean	11–12 years Male (66) Mean	11–12 years Female (56) Mean	13–14 years Male (61) Mean	13–14 years Female (57) Mean	16 years Male (38) Mean	16 years Female (38) Mean	17–18 years Male (37) Mean	17–18 years Female (29) Mean
Forehead	4.9	4.7	5.1	5.1	5.4	5.5	5.8[a]	5.4	5.8	5.5
Glabella	6.1	5.9	6.2	6.1	6.8	6.3	7.0[a]	6.1	6.4	6.1
Nasion	9.3	8.5	9.6[a]	8.8	9.5[a]	8.6	9.7[a]	8.6	9.1	8.4
Midnasal	4.2	3.9	4.3	3.9	4.5[a]	4.0	4.8[a]	4.0	4.6[a]	3.7
Rhinion	2.5	2.5	2.5	2.4	2.6	2.5	2.7	2.4	2.8	2.2
Midphiltrum	13.8[a]	12.5	14.8[a]	13.7	16[a]	14.4	17.8[a]	14.4	17.2[a]	15.2
Upper lip	12.5[a]	11.2	12.4[a]	11.4	13.1[a]	12.0	14.4[a]	11.7	14[a]	12.6
Lower lip	15.6[a]	13.7	15.6[a]	14.3	16.9[a]	15.1	17.9[a]	15.6	17.9[a]	15.6
Labiomental	10.3	9.8	10.7	9.9	11.6	10.5	12.6[a]	10.9	12.5[a]	11.1
Mental	11.7	10.8	12.2	11.3	13.1	11.9	13.7	12.2	13.7	11.8
Gnathion	7.9	7.6	8.4	8.3	8.9	8.7	9.0	8.4	9.7	8.2

[a] Significantly thicker by sex.
Note: modified from Garlie and Saunders (1999).

Table 8.3 Radiographic facial tissues measurements (mm) from North American children.

	Smith and Buschang (2001) Cephalometric White Canadians						Williamson et al. (2002) Craniographic Black North Americans											
	6–9 Years				10–15 Years				16–19 Years				7–9 Years				10–12 Years				13–15 Years			
	Male		Female		Male		Female		Male		Female		Male (n = 30)		Female (n = 32)		Male (n = 15)		Female (n = 23)		Male (n = 42)		Female (n = 18)	
Facial points	Mean	n	Mean	n	Mean	n	Mean	n	Mean	n	Mean	n	Mean	SD	Mean	SD	Mean	SD	Mean	SD	Mean	SD	Mean	SD
Forehead													5.3	1.08	5.1	1.13	5.2	0.84	5.5	1.25	6.0	1.29	5.2	0.95
Glabella	4.9	247	4.8	335	5.2	595	5.3	590	5.7	70	5.6	59	6.3	1.39	5.9	1.23	6.3	1.06	6.4	1.42	6.8	1.57	6.1	1.29
Nasion	7.6[a]	248	7.2	338	7.3	691	7.1	636	8.6[a]	186	7.4	99	6.5	1.53	6.0	1.54	6.4	1.67	6.2	1.21	6.9	1.82	6.1	1.25
Midnasal													3.6	1.04	3.7	0.88	3.6	0.81	3.9	0.87	4.4	0.82	3.9	1.04
Rhinion													3.0	0.45	3.0	0.72	3.0	0.56	3.3	0.77	3.5	0.56	2.9	0.30
Midphiltrum	12.5[a]	211	12.0	308	15.4[a]	628	14.5	622	17.7[a]	139	14.9	91	13.4	2.51	12.5	1.66	13.9	2.42	14.0	1.56	16.0	1.96	14.6	2.17
Upper lip	13.9[a]	211	12.7	303	14.8[a]	627	13.5	612	16.5[a]	135	13.5	91	12.3	1.65	11.3	1.34	12.5[a]	1.59	11.9	1.55	13.5[a]	1.69	11.5	1.17
Lower lip	15.7[a]	171	14.4	258	16.7[a]	516	15.4	489	16.9[a]	138	15.5	77												
Labiomental	10.7[a]	171	9.5	252	11.6[a]	512	10.6	488	12.4[a]	138	11.0	74												
Mental	11.5	232	11.3	327	12.3[a]	662	11.7	631	13.5[a]	164	12.3	96	10.1	2.37	10.4	2.46	11.6	2.79	11.4	2.55	13.1	3.17	12.2	2.58
Gnathion													6.7	1.89	7.4	1.84	8.1	2.41	8.2	2.04	8.3	2.72	9.1	2.59
Menton													9.2	4.07	8.4	3.50	9.2	2.38	9.7	2.90	10.2	4.14	10.0	2.56

[a] Significantly thicker by sex.

Note: SD = Standard deviation.

Table 8.4 Ultrasonic facial tissue measurements (mm) from White American children.

	4–15 years White North Americans					
	Male (n = 28)			Female (n = 22)		
Facial points	Mean	SD	Range	Mean	SD	Range
Forehead	5.7	0.59	4.6–6.7	5.5	0.46	4.7–6.6
Glabella	5.4	0.83	4–8.1	5.3	0.70	4.3–6.8
Nasion	5.3	0.81	3.6–6.6	5.3	0.95	3.5–6.8
Rhinion	2.2	0.72	1.3–3.6	2.2	0.59	1.5–3.8
Midphiltrum	10.8[a]	1.10	8.6–13.6	10.0	1.51	7.3–12.5
Upper lip	9.7	1.26	7.4–11.6	9.8	1.32	7.5–12.2
Lower lip	10.8	1.15	8.3–13.3	10.4	1.09	8.2–12.8
Labiomental	8.5	1.11	6.4–10.6	7.9	0.97	5.8–9.8
Mental tubercle	6.9	1.18	4.6–9.2	6.6	1.07	4.6–8.6
Gnathion	4.3	0.86	2.3–6.1	4.1	0.71	2.5–5.1
Frontal eminence	5.9	0.67	4.5–7.5	5.6	0.69	4.5–7
Supraorbital	6.4	1.28	4.5–9.3	6.4	0.80	5.2–8
Infraorbital	6.5	0.94	5.2–8.4	6.6	0.99	4.8–9.3
Inferior malar	17.5	2.14	14–21.4	17.1	1.76	12–20
Upper 2nd molar	12.4	1.60	8.8–14.9	12.3	1.20	10.2–14.5
Lower 2nd molar	10.7	1.69	7.4–14	10.2	1.27	7.1–12.4
Midtemporal	9.3	1.15	6.6–11.4	8.7	1.27	5.9–11
Zygomatic arch	7.4	0.90	5.8–9.2	7.2	0.96	5.5–9.3
Occlusal line	18.4	1.78	15.6–21.6	17.8	2.30	12–22.1
Gonion	7.4	1.42	4.8–10.8	7.9	1.45	5.4–11.1

[a]Significantly thicker by sex.
Note: modified from Hodson *et al.* (1985).

rhinion and lateral orbit. White girls had thicker tissue than women at the midphiltrum and supracanine points, and White boys had thicker tissues than men at the zygomatic attachment point (see Table 8.9). These differences were less obvious between Black children and adults, but Black children exhibited thicker tissue at the cheeks and philtrum, relative to the rest of the face, than Black adults. Gerasimov (1971) stated that children's faces are more soft and swollen than adults, as the thickness of soft tissues is comparatively bigger in relation to the size of the bones. Another radiographic study (Dumont, 1986) suggested that adults had thicker tissue at the brow and nasal points than children. There was thinner tissue on the faces of children at the subnasale

Table 8.5 Ultrasonic facial tissue measurements (mm) from Black North American children aged 3–18 years.

	3–8 years						9–13 years						14–18 years					
	Female (n = 52)			Male (n = 37)			Female (n = 59)			Male (n = 62)			Female (n = 25)			Male (n = 12)		
Facial points	Mean	SD	Range	Mean	SD	Range	Mean	SD	Range	Mean	SD	Range	Mean	SD	Range	Mean	SD	Range
Glabella	4.0	0.91	2–6	4.1	0.74	3–6	4.3	0.83	3–6	4.5	0.97	3–7	4.7	1.14	3–7	5.3	0.78	4–7
Nasion	4.9	0.96	3–8	5.4[a]	0.96	3–7	5.4	1.00	3–7	5.4	0.98	3–8	5.3	1.11	4–8	6.1[a]	0.51	5–7
Rhinion	1.7	0.61	1–3	1.8	0.48	1–3	1.7	0.56	1–3	1.9	0.46	1–3	1.7	0.54	1–3	2.1[a]	0.51	1–3
Lateral nostril	7.0	1.48	5–11	7.3	1.68	5–11	7.6	1.58	5–12	7.4	1.91	4–13	8.1	2.14	5–12	7.9	1.98	5–10
Midphiltrum	8.9	1.57	6–14	9.0	1.18	6–11	9.6	1.56	7–13	10.0[a]	1.69	7–18	9.9	2.20	7–16	12.1[a]	1.73	10–15
Labiomental	8.2	2.05	3–15	8.6	1.44	6–12	10.3	1.77	7–15	9.8	1.84	6–13	10.1	1.79	7–13	12.6[a]	1.93	10–16
Mental	8.3	2.16	4–14	8.3	1.59	6–11	10.0	2.60	5–16	9.9	3.03	5–18	10.0	2.65	4–15	9.5	2.78	5–13
Menton	4.8	1.61	2–10	4.5	1.12	2–6	5.8	2.15	2–12	5.5	2.09	2–11	5.6	1.93	2–10	6.3	1.86	4–10
Supraorbital	4.5	1.02	3–7	4.5	0.65	3–6	5.3	1.03	3–8	5.2	1.12	3–9	5.7	1.46	4–10	5.8	0.94	4–7
Infraorbital	5.6	1.14	3–9	5.6	1.07	3–8	6.1	1.12	4–10	5.8	1.19	3–9	6.4	1.50	4–11	6.0	0.74	5–7
Supracanine	8.8	1.59	5–14	8.9	1.86	6–15	10.0	1.79	7–16	10.7	2.74	7–27	10.6	1.50	8–13	12.3	2.05	9–17
Subcanine	9.0	2.20	5–15	8.5	1.24	6–11	10.2	2.16	6–17	11.0	3.02	7–24	11.0	2.25	7–16	12.8	2.67	8–17
Upper 2nd molar	23[a]	3.39	15–32	22.1	2.47	17–27	24.5[a]	3.72	18–34	23.6	4.35	12–33	27.6[a]	3.52	22–37	26.0	2.89	21–30
Lower 2nd molar	18.0	3.26	10–26	17.4	2.68	10–25	20.0	3.58	10–26	20.1	4.18	11–28	23.2	3.99	18–33	21.9	4.91	12–29
Midmandible	9.8	3.16	5–20	8.7	2.03	5–14	10.8	2.99	6–18	10.3	3.86	4–20	12.0	3.16	7–20	11.2	3.93	7–20
Lateral orbit	3.9	0.89	2–6	4.1	0.85	2–6	4.4	1.24	2–10	4.4	0.89	3–7	4.6	1.08	3–8	4.4	0.67	3–5
Zygomatic attach	8.3[a]	2.23	4–15	7.8	1.55	5–12	8.9[a]	2.22	6–14	8.3	2.66	4–15	9.2[a]	1.68	6–13	7.3	2.05	5–12
Gonion	13.5	2.87	8–21	12.8	2.02	10–17	14.6	3.41	3–23	14.7	3.06	9–22	16.2	3.36	10–23	17.9	3.63	11–24
Root Zygoma	4.7	1.21	3–8	4.2	0.98	3–6	4.8	1.55	3–8	5.0	1.73	2–12	6.2	2.30	3–13	6.0	2.37	3–11

[a] Significantly thicker by sex.
Note: modified from Manhein *et al.* (2000).

Table 8.6 Ultrasonic facial tissue measurements (mm) from White North American children aged 3–18 years.

	3–8 years						9–13 years						14–18 years					
	Female (n = 43)			Male (n = 36)			Female (n = 51)			Male (n = 45)			Female (n = 35)			Male (n = 27)		
Points	Mean	SD	Range	Mean	SD	Range	Mean	SD	Range	Mean	SD	Range	Mean	SD	Range	Mean	SD	Range
Glabella	3.9	0.98	2–7	4.0	0.84	3–6	4.4	1.08	2–7	4.6	1.04	2–7	4.6	0.98	3–6	5.0	0.73	4–7
Nasion	5.0	0.94	3–7	5.7[a]	0.96	3–8	5.5	1.03	3–8	5.7	1.09	3–8	5.4	0.88	4–8	6.3[a]	1.07	4–8
Rhinion	1.7	0.52	1–3	1.8	0.67	1–4	1.5	0.54	1–3	1.6	0.53	1–3	1.8	0.51	1–3	2.0	0.44	1–3
Lateral nostril	7.0	1.86	4–12	7.2	1.75	4–11	7.7	2.00	4–15	7.4	1.71	4–15	7.7	1.78	5–12	7.8	1.96	5–12
Midphiltrum	8.3	1.35	6–12	9.0[a]	1.59	6–12	9.4	1.54	6–13	9.7[a]	1.50	6–13	9.4	1.46	7–12	11.2[a]	1.98	7–15
Labiomental	7.6	1.51	5–12	8.1[a]	1.79	6–12	9.0	1.45	6–13	9.6[a]	1.75	6–13	9.7	1.25	8–13	10.4[a]	1.28	7–13
Mental	7.4	1.81	4–11	8.3	2.14	4–12	8.8	1.98	5–14	8.7	2.93	5–14	8.7	1.75	5–14	9.3	1.90	7–14
Menton	4.2	1.19	2–8	4.6	1.13	3–7	5.5	1.64	2–11	5.5	1.44	2–11	5.5	1.36	4–9	6.0	1.57	4–11
Supraorbital	4.4	1.15	3–7	4.6	0.84	3–6	5.1	0.92	3–8	5.2	0.82	3–8	5.7	1.47	4–12	5.7	0.83	4–7
Infraorbital	5.6	1.12	3–8	5.5	0.94	4–8	5.6	1.08	4–8	5.9	1.14	4–8	6.0	1.25	3–9	5.3	1.23	4–9
Supracanine	8.4	1.29	6–11	9.4[a]	1.98	6–14	9.8	1.68	7–14	10.0	1.77	7–14	10.3	3.22	7–26	11.7[a]	2.33	8–19
Subcanine	7.9	1.44	5–11	8.4	1.40	6–13	9.2	1.61	6–13	9.6	1.70	6–13	9.8	2.40	6–21	10.6	2.32	7–17
Upper 2nd molar	22.7	3.48	14–30	23.3	3.73	14–31	24.3	2.88	19–32	24.7	4.30	19–32	26.8	4.96	5–34	27.4	3.38	22–35
Lower 2nd molar	18.9	3.59	8–24	20.7	3.64	13–31	20.8	3.63	13–29	21.6	3.71	13–29	23.2	4.58	5–30	23.2	3.48	15–31
Midmandible	10.5	3.33	4–18	10.4	2.80	6–15	11.7	3.24	4–18	12.1	3.99	4–18	13.4	2.76	9–19	12.3	4.49	6–24
Lateral orbit	4.0	0.89	3–6	4.1	0.91	2–6	4.6	1.09	3–9	4.4	0.87	3–9	4.5	0.85	3–6	4.3	0.86	3–7
Zygomatic attach	8.4	2.44	5–15	8.4	2.29	5–15	9.5	2.24	5–14	9.1	2.46	5–14	9.5	1.85	6–16	8.0	1.76	6–13
Gonion	13.9	3.27	7–22	13.7	2.89	8–20	14.4	2.90	8–19	15.4	3.63	8–19	17.0	2.67	12–22	18.1	3.04	14–24
Root Zygoma	4.6	1.51	3–10	4.8	1.02	3–7	5.2	1.58	3–10	5.4	1.79	3–10	6.8	1.88	4–12	6.0	2.07	4–12

[a] Significantly thicker by sex.

Note: modified from Manhein *et al*. (2000).

Table 8.7 Ultrasonic facial tissue measurements (mm) from Hispanic North American children aged 3–18 years.

| | 3–8 years | | | | | | 9–13 years | | | | | | 14–18 years | | | | | |
| | Female (n = 6) | | | Male (n = 3) | | | Female (n = 9) | | | Male (n = 8) | | | Female (n = 1) | | | Male (n = 4) | | |
Facial points	Mean	SD	Range	Mean	SD	Range	Mean	SD	Range	Mean	SD	Range	Mean	SD	Range	Mean	SD	Range
Glabella	4.2	0.75	3–5	4.7	0.58	4–5	3.8	0.83	3–5	4.1	0.83	3–5	7.0			4.5	1.00	4–6
Nasion	5.0	1.10	3–6	6.3	1.15	5–7	5.3	0.87	4–6	4.9	1.35	3–7	5.0			4.8	0.50	4–5
Rhinion	1.7	0.52	1–2	1.7	0.58	1–2	1.6	0.53	1–2	1.6	0.52	1–2	1.0			1.5	0.58	1–2
Lateral nostril	6.3	1.03	5–8	6.3	1.53	5–8	5.7	1.12	5–8	7.9	2.23	5–12	9.0			5.0	0.82	4–6
Midphiltrum	8.0	1.55	7–10	7.3	0.58	7–8	9.2	1.20	8–11	9.3	1.75	5–10	8.0			11.5	1.29	10–13
Labiomental	8.7	2.07	6–11	7.0	2.00	5–9	9.2	1.48	7–12	10.0	1.85	6–12	11.0			11.3	2.06	9–14
Mental	8.0	2.00	5–10	6.0	1.00	5–7	8.4	1.59	6–11	8.4	2.77	5–13	15.0			10.3	0.96	9–11
Menton	4.2	1.72	2–6	4.7	1.53	3–6	5.1	1.36	3–7	5.1	0.99	4–6	9.0			5.8	0.96	5–7
Supraorbital	4.2	0.75	3–5	4.3	0.58	4–5	4.9	0.93	3–6	4.9	0.99	4–6	7.0			5.5	1.29	4–7
Infraorbital	5.5	1.87	3–8	5.0	2.00	3–7	5.0	1.12	3–6	6.4	1.41	4–9	10.0			5.8	0.96	5–7
Supracanine	9.3	2.66	7–14	8.0	1.00	7–9	10.3	1.66	9–13	10.0	2.33	6–13	11.0			12.0	0.82	11–13
Subcanine	8.2	2.32	6–11	6.7	0.58	6–7	8.3	1.32	6–10	10.8	2.12	8–14	10.0			10.0	3.16	6–13
Upper 2nd molar	24.8	3.37	20–28	19.7	3.51	16–23	24.6	4.13	16–29	24.4	2.33	21–28	32.0			25.3	4.27	19–28
Lower 2nd molar	20.8	6.15	10–28	14.7	4.73	11–20	20.0	5.12	9–26	21.8	2.83	18–27	24.0			21.0	1.41	20–23
Midmandible	11.5	3.94	5–16	7.3	4.04	5–12	11.3	2.78	6–15	10.8	3.11	5–14	18.0			10.3	4.57	5–15
Lateral orbit	4.3	0.82	3–5	3.0	0.00	3–3	3.8	0.44	3–4	4.6	0.52	4–5	5.0			4.3	0.96	3–5
Zygomatic attach	8.5	2.66	5–13	6.3	2.10	4–8	7.4	1.13	6–9	8.4	1.69	6–11	14.0			7.8	1.89	5–9
Gonion	14.0	3.41	8–18	13.7	5.03	9–19	14.6	3.05	10–19	15.4	4.63	7–21	24.0			15.3	4.86	9–20
Root Zygoma	4.3	0.82	3–5	4.3	2.31	3–7	4.6	1.33	3–6	6.3	1.28	5–8	8.0			4.8	1.50	3–6

Note: modified from Manhein *et al.* (2000).

Table 8.8 Ultrasonic facial tissue measurements (mm) from White British children aged 11–18 years.

| | 11–12 years | | | | | | 13–14 years | | | | | |
| | Male (n = 30) | | | Female (n = 28) | | | Male (n = 21) | | | Female (n = 23) | | |
Facial points	Mean	SD	Range	Mean	SD	Range	Mean	SD	Range	Mean	SD	Range
Forehead	4.6	0.97	3–7	4.3	0.83	3–6	5.0	0.89	4–7	5.0	0.88	3–6
Glabella	4.9	0.89	4–7	4.5	0.63	3–6	5.2	0.89	4–7	5.2	0.86	4–7
Nasion	4.7	0.88	3–7	4.3	0.72	3–6	5.2	0.89	4–7	4.7	0.95	3–7
Rhinion	2.7[a]	0.73	1–4	2.2	0.40	1–3	2.4	0.84	1–4	2.2	0.61	1–4
Midphiltrum	11.0	1.60	7–13	9.9	1.91	6–13	10.3	2.20	7–15	9.6	2.14	6–13
Upper lip	10.7	1.56	7–13	10.0	1.53	7–13	10.6	1.78	8–17	9.2	1.66	6–12
Lower lip	12.1	1.79	9–15	11.4	1.48	8–14	12.2[a]	2.13	5–12	10.5	1.74	6–13
Labiomental	10.2	1.54	7–14	9.8	1.44	7–13	8.9	2.09	7–13	9.4	2.29	5–14
Mental	9.8	2.08	6–16	10.2	2.10	7–15	10.1	1.98	5–8	8.9	1.94	6–12
Gnathion	6.7	1.23	5–10	6.0	1.08	4–8	6.0	0.94	7–13	6.0	1.22	4–9
Lateral forehead	5.7	1.00	4–8	4.7	0.81	3–6	5.4	0.95	4–8	5.4	0.93	4–7
Supraorbital	6.1[a]	0.96	4–8	5.4	0.90	4–8	6.6	0.98	5–8	6.5	0.91	4–8
Infraorbital	7.6[a]	1.12	6–10	7.0	1.04	5–9	8.3	1.30	6–10	7.3	1.20	5–9
Lateral nose	6.9	1.37	4–10	6.3	1.33	4–10	6.4[a]	1.79	4–10	5.9	1.25	4–10
Lateral orbit	6.9	1.08	5–9	7.4	1.97	5–15	7.1	1.12	5–8	7.4	1.15	5–10
Zygomatic attach	12.3	3.12	8–3	13.0	3.89	7–21	10.7	2.48	7–16	10.2	1.81	8–15
Upper 1st molar	16.7	5.46	10–28	17.4	6.76	10–30	14.7	3.31	10–21	16.2	3.15	11–22
Lower 1st molar	16.0	4.46	10–27	15.5	4.66	9–27	14.6	2.28	9–20	14.5	2.84	9–20
Midmandible	10.2	2.59	7–22	9.8	2.05	7–14	9.1	1.84	7–13	9.1	1.87	6–15
Zygomatic arch	8.0	1.53	5–11	8.0	2.05	6–10	8.4	1.63	6–12	7.9	1.87	5–12
Midmasseter	18.4	4.73	7–28	17.3	4.85	9–25	17.4	4.94	9–25	16.6	4.97	8–23

Facial points	15–16 years						17–18 years					
	Male (n = 30)			Female (n = 28)			Male (n = 21)			Female (n = 23)		
	Mean	SD	Range	Mean	SD	Range	Mean	SD	Range	Mean	SD	Range
Forehead	4.6	0.66	3–6	4.7	0.64	3–6	4.8	0.77	3–6	4.8	1.04	3–7
Glabella	5.0	0.89	4–6	4.9	0.66	4–7	5.2	0.74	4–7	5.2	1.20	3–8
Nasion	4.9	1.03	3–7	4.8	0.76	3–7	5.3	0.87	2–9	5.0	1.31	2–9
Rhinion	2.5	0.66	1–4	2.6	0.62	2–4	2.5	0.61	2–4	2.5	0.82	1–4
Midphiltrum	10.9	1.97	6–14	11.2	1.83	7–16	12.5[a]	1.86	9–16	10.9	1.49	7–13
Upper lip	11.6	1.42	8–14	11.2	1.54	8–15	12.1	1.85	9–18	10.8	1.62	8–14
Lower lip	12.8	2.19	9–19	13.0	2.03	8–17	13.8[a]	2.40	9–20	12.0	1.59	9–16
Labiomental	10.8	2.12	6–14	11.1	1.65	7–13	11.9	1.30	9–16	11.1	1.02	9–13
Mental	10.9	1.40	8–13	10.0	1.84	6–13	10.9	0.97	9–13	11.0	1.92	7–15
Gnathion	6.9	1.48	4–11	7.2	1.18	5–9	7.7	1.06	6–10	7.3	1.39	5–10
Lateral forehead	5.6	0.76	4–7	5.6	0.95	4–8	5.2	1.23	4–8	5.3	1.34	3–8
Supraorbital	6.4	0.73	5–8	6.0	0.90	4–8	5.9	1.23	4–8	6.2	1.16	4–8
Infraorbital	7.2	0.69	6–9	7.8	1.61	5–10	7.9	1.17	6–11	7.5	1.21	4–10
Lateral nose	6.1	0.96	5–9	6.9	1.51	5–10	6.7	1.52	5–10	6.9	1.35	4–10
Lateral orbit	7.2	1.28	5–10	7.5	1.36	5–10	7.4	1.16	5–9	7.8	1.52	4–11
Zygomatic attach	11.5	2.27	7–15	13.6	2.71	9–20	11.3	1.79	7–14	13.2	2.33	9–19
Upper 1st molar	18.4	4.75	12–25	20.9[a]	3.41	14–27	18.3	4.83	10–27	19.2[a]	5.32	9–28
Lower 1st molar	16.6	4.33	11–25	18.4	4.26	12–29	16.7	3.36	12–24	17.0	4.41	10–27
Midmandible	9.4	1.90	6–14	10.9	3.06	8–22	9.3	2.38	6–16	9.6	2.19	6–13
Zygomatic arch	7.5	1.05	6–10	8.3	3.06	6–10	7.9	1.28	6–11	8.4	2.19	5–12
Midmasseter	18.8	4.50	8–24	20.2	4.04	11–25	21.9	3.37	11–24	21.5	3.91	11–31

[a] Significantly thicker by sex.
Source: Wilkinson (2002).

and beneath the chin. Girls had thicker tissue at the upper lip, and boys had thinner tissue at the labiomental point, than adults. A comparison was made between White European adults (Helmer, 1984) and children (Wilkinson, 2002) (see Table 8.10 and Fig. 8.6). Juvenile faces were generally smaller than adult faces, and this was reflected in the smaller tissue depths at the majority of facial points. Juvenile faces had thicker tissues than adult faces at the infraorbital, lateral orbit and zygomatic arch points. In addition, girls had thicker tissue than women at the masseter and lateral nasal points.

In conclusion, the child's face tends to be relatively larger at the cheek area and over the bridge of the nose. The child has fuller cheeks and a smoother jawline than the adult, for both males and females.

Differences in juvenile tissue depths between sexes

Those studies that did not distinguish between different age groups found similar tissue measurements between boys and girls. Altemus (1963) and Heglar and Parks (1980) studied children from the ages of 13–16 years and 10–18 years respectively and found that, although boys appeared to have thicker mean tissue depths than girls at the majority of points, the variation in measurements at each point was enormous (see Table 8.1). Another study (Hodson et al., 1985; see Table 8.4) measured children between the ages of four and 15 years, and found that girls had significantly thinner tissue at the midphiltrum, but that there were no other significant sex differences in the facial tissues. However, the ranges of measurement at all points were wide. They also suggested that midphiltral tissue increased with age in females, the labiomental tissue increased with age in males, and the frontal eminence tissue increased with age in both sexes. A decrease in tissue depth at the infraorbital, inferior malar, upper molar and gonial points (the cheek region) was shown with an increase in age in both girls and boys. Another study (Dumont, 1986) found no significant differences in midfacial tissue depths between boys and girls at ages

Table 8.9 Comparison of facial tissue measurements (mm) between White and Black American adults and children.

Facial points	White adults		White children		Black adults		Black children	
	Male (48) Mean	Female (82) Mean	Male (108) Mean	Female (129) Mean	Male (22) Mean	Female (44) Mean	Male (111) Mean	Female (136) Mean
Glabella	5.5[a]	4.9	4.5	4.3	5.3	4.6	4.6	4.3
Nasion	6.6[a]	5.8	5.9	5.3	6.2	5.7	5.6	5.2
Rhinion	1.6	1.8	1.8	1.7	2.0	1.7	1.9	1.7
Lateral nostril	9.6[a]	9.3[a]	7.5	7.5	9.8[a]	8.4[a]	7.5	7.6
Midphiltrum	10.0	8.1	10.0	9[a]	12[a]	8.7	10.2	9.5[a]
Labiomental	12[a]	10.3[a]	9.4	8.8	12.7[a]	11.2[a]	10.3	9.5
Mental	11.2[a]	10.4[a]	8.8	8.3	12.2[a]	10.9[a]	9.2	9.4
Menton	7[a]	6.5[a]	5.4	5.1	7.9[a]	6.8[a]	5.3	5.4
Supraorbital	6.1[a]	6[a]	5.2	5.1	6.4	6.0	5.2	5.2
Infraorbital	6.0	6.5[a]	5.6	5.7	6.4	6.3	5.8	6.0
Supracanine	10.3	8.2	10.4	9.5[a]	11.6[a]	9.5	10.6	9.8
Subcanine	10.9[a]	9.2	9.5	9.0	12.6[a]	11.6[a]	10.8	10.1
Upper 2nd molar	28.9[a]	27.1[a]	25.1	24.6	27.8[a]	26.7[a]	23.9	25.0
Lower 2nd molar	22.1	23.1[a]	21.8	21.0	24.1[a]	21.8[a]	19.8	20.4
Midmandible	14.3[a]	14.2[a]	11.6	11.9	13.7[a]	13[a]	10.1	7.2
Lateral orbit	4.8	4.6	4.3	4.4	4.3	4.9	4.2	4.3
Zygomatic attach	7.7	9.8[a]	8.5[a]	9.1	7.4	9.9	7.8	8.8
Gonion	18.2[a]	16.1[a]	15.7	15.1	20.9[a]	16[a]	15.1	14.8
Root Zygoma	6.3[a]	6.4[a]	5.4	5.5	6.6[a]	6.0	5.1	5.2

[a] Significantly thicker between adults and children.
Modified from Manhein *et al.* (2000).

Table 8.10 Comparison of facial tissue measurements (mm) between White European adults and children.

| | White adults (Helmer, 1984) 30–39 years | | | | White children (Wilkinson, 2002) 11–18 years | | | |
| | Male (14) | | Female (13) | | Male (99) | | Female (101) | |
Facial points	Mean	Range	Mean	Range	Mean	SD	Mean	SD
Forehead	5.0	4.5–5.5	4.5	4.3–5	4.7	0.85	4.6	0.88
Glabella	6.2[a]	5.3–6.5	5.7	5.2–6	5.1	0.84	4.9	0.89
Nasion	7.3[a]	7–7.5	6.5[a]	6–6.8	5.0	0.94	4.7	0.98
Rhinion	2.5	2–2.7	2.5	2.2–2.7	2.6	0.71	2.4	0.64
Midphiltrum	14.6[a]	13.5–16	12.8[a]	12.2–13.5	11.2	2.02	10.4	1.94
Upper lip	12.3	10.5–12.7	10.7	9.8–11	11.2	1.74	10.3	1.72
Lower lip	14.9	13.7–15.3	12.0	11.3–12.5	12.7	2.20	11.7	1.92
Labiomental	12.1	11.5–12.8	10.8	10–12	10.5	2.03	10.3	1.78
Mental	10.3	9.2–12	10.0	8.8–10.5	10.4	1.73	10.0	2.07
Gnathion	8.3[a]	7–9	7.2	6.2–8.2	6.9	1.32	6.6	1.34
Lateral forehead	6.0	5.5–6.5	5.0	5–5.5	5.5	1.01	5.2	1.05
Supraorbital	7.3[a]	6.5–7.5	6.5	6–6.7	6.2	1.01	6.0	1.05
Infraorbital	5.0	4.5–5.5	5.5	4.8–6	7.7[a]	1.15	7.4[a]	1.30
Lateral nose	7.4	6.7–8.5	6.3	5.2–7.5	6.5	1.44	6.5	1.41
Lateral orbit	5.2	5–5.5	5.2	4.5–5.5	7.4[a]	1.15	7.1[a]	1.55
Zygomatic attach	9.9	8.5–10.5	10.3	9.8–11.7	11.5[a]	2.53	12.6[a]	3.10
Upper 1st molar	22.0[a]	19–24	21.5	19.5–23.5	17.1	4.90	18.5	5.20
Lower 1st molar	18.5	16.7–21.7	19.0	17–22.8	16.0	3.81	16.3	4.34
Midmandible	11.9[a]	10–14	11.5	10–12.2	9.5	2.25	9.8	2.41
Zygomatic arch	5.3	4.5–6.3	5.2	4.5–5.3	8.0[a]	1.40	8.2[a]	2.41
Midmasseter	21.3	18.7–22.2	18.3	15.8–19.2	19.2	4.65	18.9	4.82

[a]Significantly thicker between adults and children.

nine to 11 years. Boys had thicker tissue than girls in the 12–15 years age group, with the greatest differences at the nasal, upper lip and labiomental regions (see Table 8.1). Dumont suggested that the onset of puberty leads to a divergence at these points. As Table 8.2 shows, Garlie and Saunders (1999) studied 615 children, aged between eight and 18 years, and found that boys generally have greater mean tissue thicknesses than girls. Boys were shown to have thicker tissues than girls at the glabella, nasion, nasal, midphiltrum, upper lip, lower lip and labiomental points, but these differences became more significant after the age of 14 years. They also showed that there was an increase in tissue thickness as individuals increased in age, but this correlation was weak

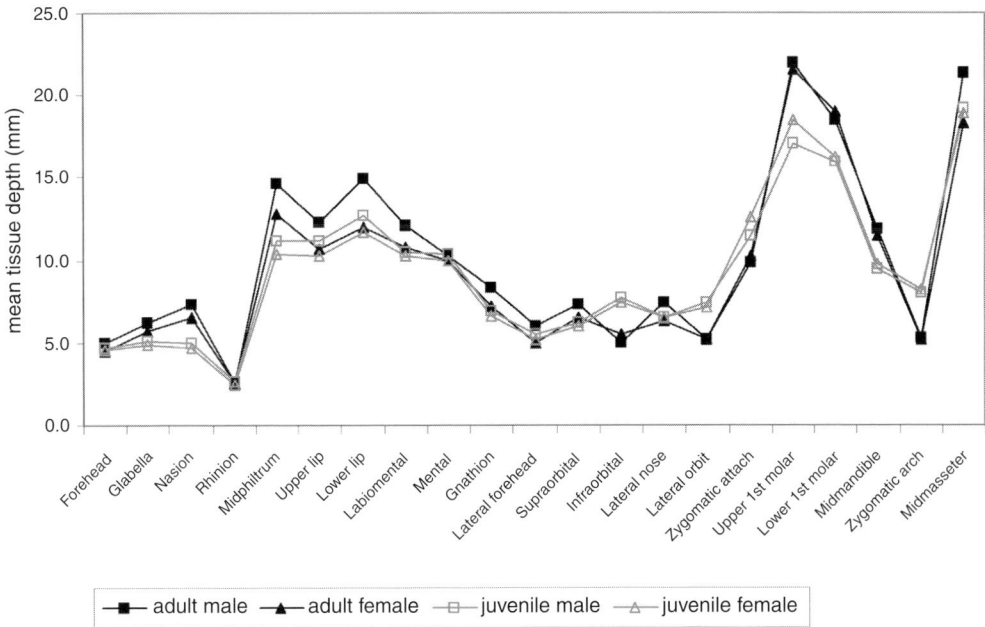

Fig. 8.6 Comparison of facial tissue measurements (mm) between White European adults (Helmer, 1984) and children (Wilkinson, 2002).

and other factors were indicated. They suggested that similar tissue data be used for the facial reconstruction of young boys and girls, but that separate standards may be necessary from the time of puberty, as there appeared to be a divergence in facial tissue thickness between boys and girls at the adolescent growth spurt.

A further radiographic study (Smith & Buschang, 2001; see Table 8.3) found that the nasion and upper lip tissues were significantly thicker for boys than for girls by age six years, lower lip tissues by age seven years, labiomental tissues by age nine years, midphiltral tissues by age 12 years and mental tissues by age 15 years. They found that forehead tissue changed little with age, and midfacial regions had the greatest change with age. They also found wide variation in measurements at all points, and this was not explained by age or sex. Another juvenile craniographic study agreed with these conclusions (Williamson *et al.*, 2002; see Table 8.3), and found that there was a general increase in tissue with age in boys and girls; but that the growth, rather than being slow and gradual,

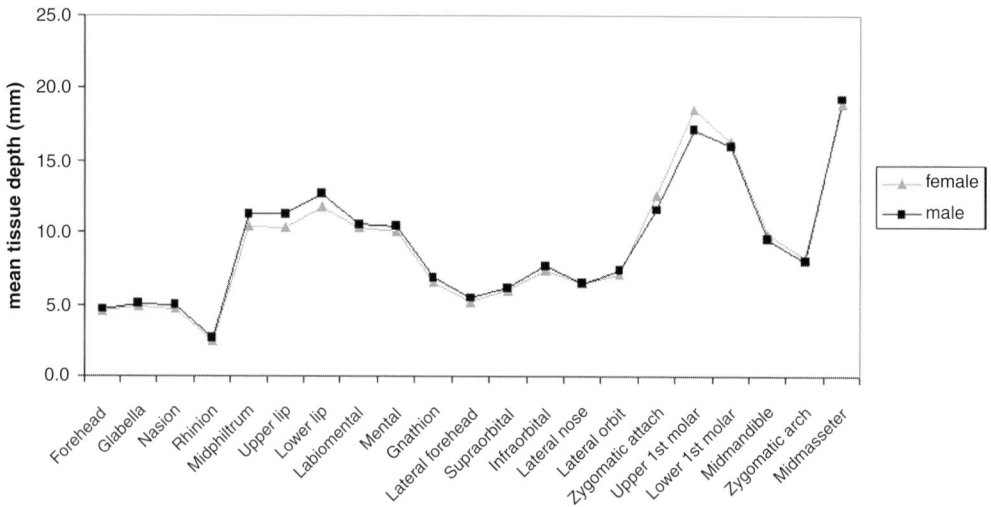

Fig. 8.7 Comparison of facial tissue measurements (mm) between sexes for White European children (Wilkinson, 2002).

occurs in spurts. They found that upper lip and eye tissues have a growth spurt at the age of 13 years, and chin tissues exhibit a significant increase at approximately nine years of age. They found significantly thicker tissue in boys than girls at the upper lip point only. The ultrasonic imaging techniques of tissue depth measurement confirmed these results and gave further information regarding the cheeks and jawline. Manhein *et al.* (2000) found that White and Black boys have consistently larger tissue depths than the girls at the midphiltrum, nasion and cheek points. They found no significant differences between tissues for Hispanic boys and girls. They found that there was an increase in tissue depth with age at most points for boys and girls, except at the nasal and zygomatic attachment points (see Tables 8.5–8.7). The study of British children (Wilkinson, 2002) suggested that the tissues at the midphiltrum, upper lip and lower lip points were consistently thicker in the boys than the girls. Tissues were consistently thicker at the zygomatic attachment point in the girls than the boys (see Table 8.8 and Fig. 8.7). Boys of all ages had more tissue at the brow and lips, and girls of all ages had fatter cheeks. The buccal fat pads are most apparent during childhood and especially in young prepubescent girls. The postpubertal females are more likely to show larger tissue thicknesses at the cheeks, since females retain more fatty tissue over the surface of the facial musculature, even in adulthood. Boys showed a general increase in tissue depth with an increase in age at the midline and the cheek points. The girls showed increased tissue depth with age at all the points, except the infraorbital, lateral orbit, midzygomatic arch and midmandibular points. Wilkinson suggested that the majority of facial changes associated with puberty occurs in boys in the 11–16 year age group and in girls in the 11–14 year age group. Boys differ from girls in that they have a growth spurt in nasal height at puberty, and they also show a growth spurt in soft tissue on the chin at puberty. In contrast, the girls have a decrease in growth of chin tissue after the age of nine years. The adolescent growth spurt leads to a great enlargement at the mandible, which is largest in boys, resulting in more projecting and robust jaws. This may explain the increase in tissue depth with age at the jaw, which may accompany this skeletal development.

The studies for British children (Wilkinson, 2002) and American children (Manhein et al., 2000) indicated that the facial points that showed the greatest standard deviation were the zygomatic attachment, upper molar, lower molar and midmasseter points, and these were consistently high for all age groups (see Tables 8.5–8.8). The girls had greater standard deviations than the boys at all the cheek points. This may be due to the earlier onset of puberty in females, since the greatest female facial changes occur at the cheek region. There was wide variation in the cheek tissues for all the children, which may be due to the wide variation in the onset of puberty. Wilkinson (2002) found that several facial points at the cheek area, including the labiomental, mental, infraorbital, zygomatic attachment, midmasseter, upper molar and lower molar points, exhibited tissue depth distributions with two peaks (see Fig. 8.8). These distributions suggested that within the sample there were two different facial types – thin and fat-faced individuals. These facial points were areas of the face where fatty tissue develops during

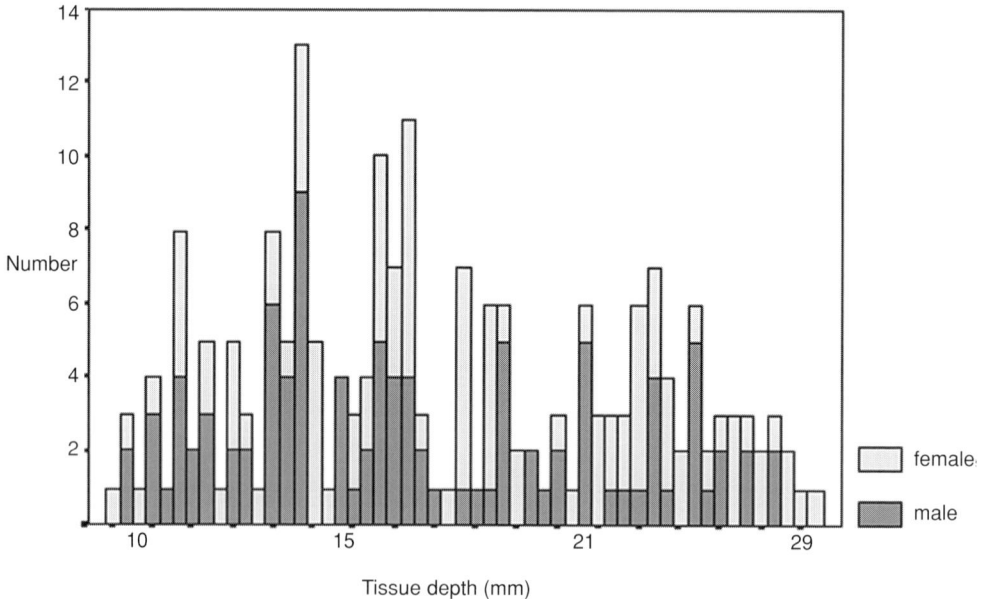

Fig. 8.8 Distribution of facial tissue measurements at the upper first molar point for White European children (Wilkinson, 2002).

childhood – namely the cheeks and chin. The suggestion that two facial types exist in this sample may be a reflection of the variation in the onset of puberty and facial maturation.

Differences in juvenile tissue depths between different ethnic groups

The White British study (Wilkinson, 2002) showed similar facial tissue distribution to the White American (Manhein *et al.*, 2000) study (see Tables 8.6 and 8.8). However, larger tissue depths were recorded in the White American than the White British children at the upper molar, lower molar and midmandibular points. The measurement variation at these points was wide in both studies. The British study showed two separate facial types (fat and thin) at the cheeks, and the fatter-cheeked group appeared to correspond more closely with the results from the American study. This may be a reflection of national or nutritional variation, as current evidence suggests that 20 per cent of American children are over-weight, as opposed to 11 per cent of British children (Anand *et al.*, 1999; BBC News, 1999). However, other studies of Black American children (Williamson *et al.*, 2002) showed consistently smaller facial tissue measurements than the equivalent ultrasonic measurements of Black Americans (Manhein *et al.*, 2000; see Tables 8.3 and 8.5).

More research needs to be carried out into facial tissue depths within this age group, since the variation between subjects appears very large, and further studies may illuminate differences with regard to facial constitution and age. Future research into different juvenile ethnic groups would be invaluable to this field.

Juvenile facial reconstruction method

The methodology of juvenile facial reconstruction is essentially the same as that for adult facial reconstruction, but there are a small number of variations. The skull of a child should not be mounted on the Frankfurt Plane, but rather with the face tilted upwards. This is the angle at which adults tend to view the faces of children, and may encourage recognition from the reconstruction (see Fig. 8.9).

The muscles of the face are modelled onto the skull, one by one, following the method outlined in Chapter 6. Muscle attachment will be less defined on juvenile skulls than adult skulls. Areas to pay particular attention to are the teeth and the eyes. The dental occlusion will provide vital information regarding the mouth shape, the projection of the lips and the position of the corners of the mouth. However, the relationship between the teeth height and the lip thickness is uncertain in children. The size of deciduous teeth may not be an accurate indication of lip thickness, as the child may be close to permanent dentition eruption, or may have both primary and secondary dentition present. A judgment on lip shape must be made based on experience, and portrayal of

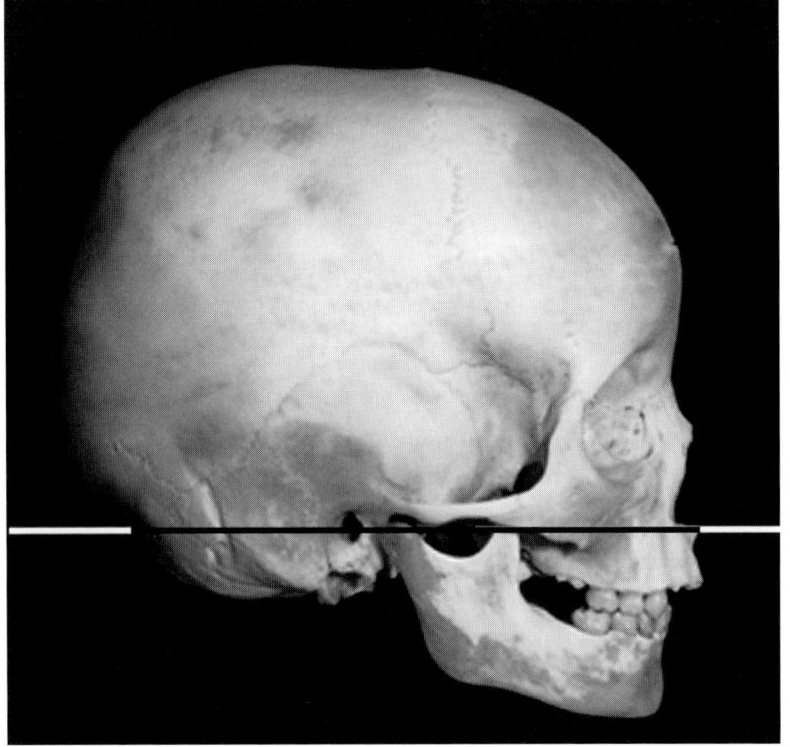

Fig. 8.9 Mounting a juvenile skull for facial reconstruction. The skull is positioned so that it appears to look upwards from the horizontal.

the lips should be in sympathy with the rest of the face. It may be necessary to illustrate the dental pattern of the child in the reconstruction, as the teeth are often vital in a child's identification. An open, but unsmiling mouth is the optimal form. The orbital depth suggests the eyeball protrusion, and this may be especially important in juvenile faces, as the eyes are large relative to the size of the face.

The faces of children tend to exhibit fuller cheeks and jawlines, due to the presence of buccal fat pads. It is preferable to model the subcutaneous fat onto the muscle structure before the addition of the skin layer. This is added as small pieces of clay, and will fill out the cheek areas prior to covering with a smooth skin (see Fig. 8.10). This will allow the shape of the cheeks to develop in a more natural

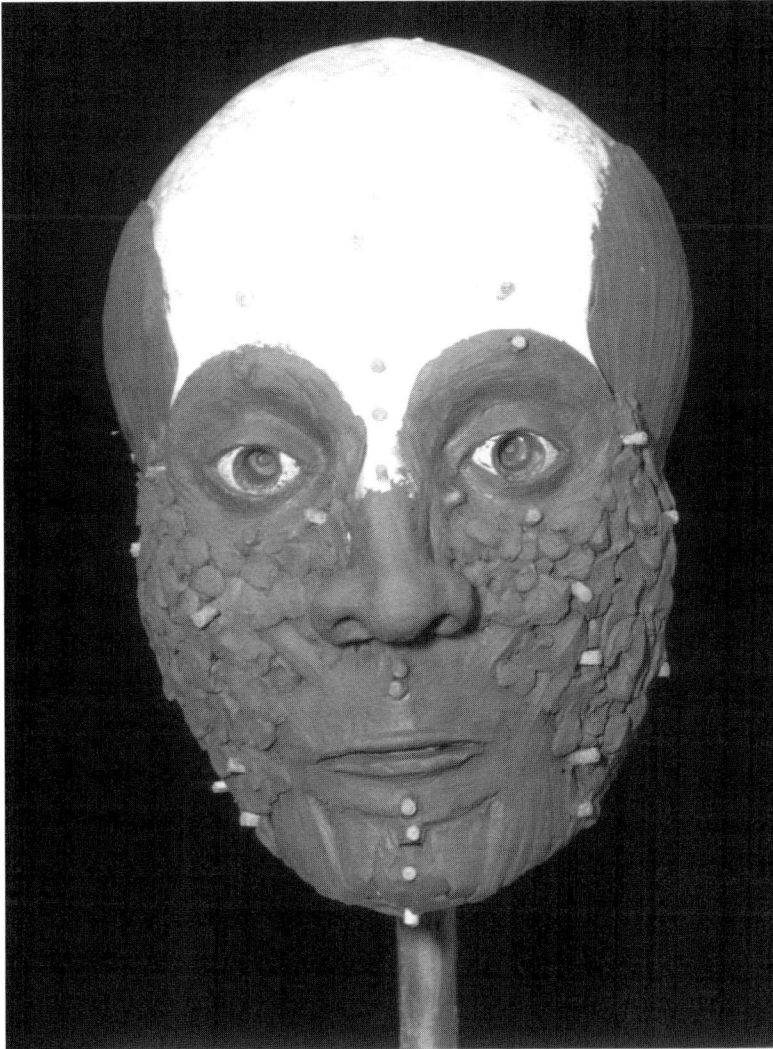

Fig. 8.10 Modelling the cheek regions of a juvenile reconstruction. Clay is placed onto the muscle structure to represent the subcutaneous fat layer.

and realistic manner than attempting to add clay onto the cheek regions following the addition of the skin layer. The skin surface of juvenile faces is smooth and unwrinkled, and the reconstruction will benefit from a realistic skin texture. The desired texture can be achieved by patting of the smooth clay surface with a soft sponge (see Fig. 8.11).

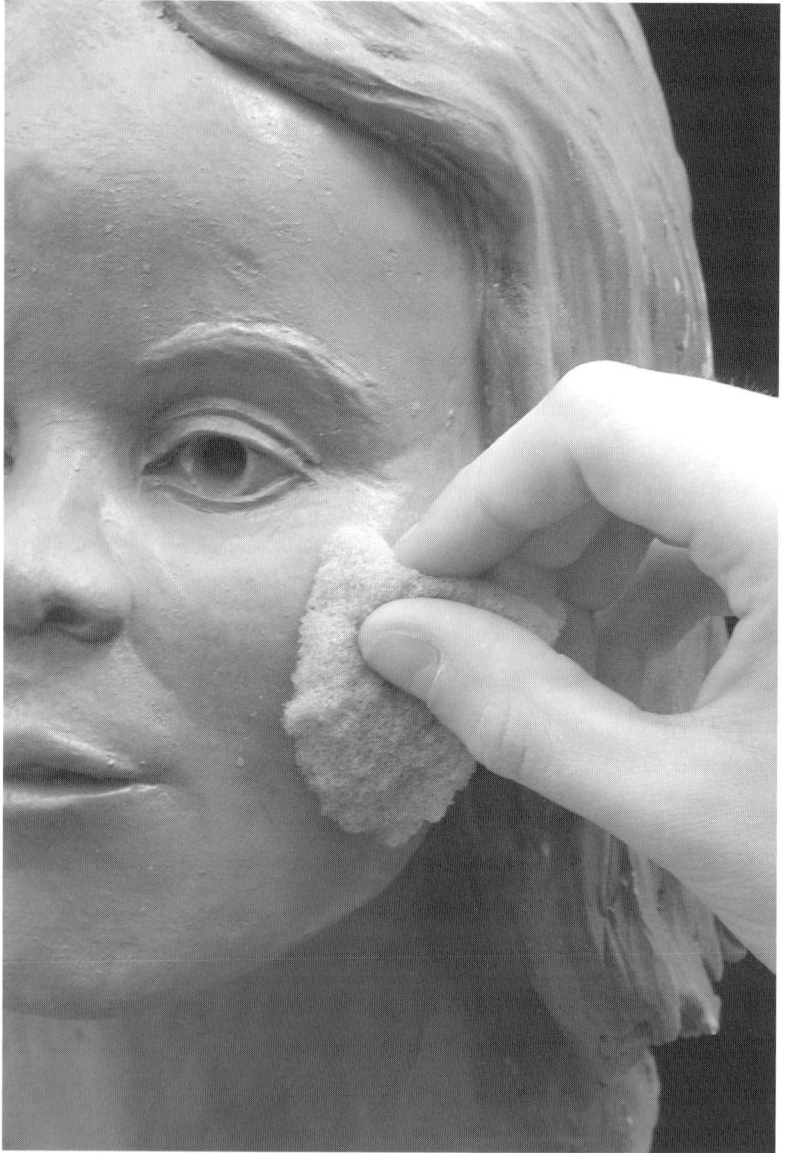

Fig. 8.11 Adding skin texture to a juvenile facial reconstruction.

The accuracy of juvenile facial reconstruction

The facial reconstruction of childrens' faces has not been stud-ied in detail, although numerous forensic investigations have employed facial reconstruction in order to provoke recognition for

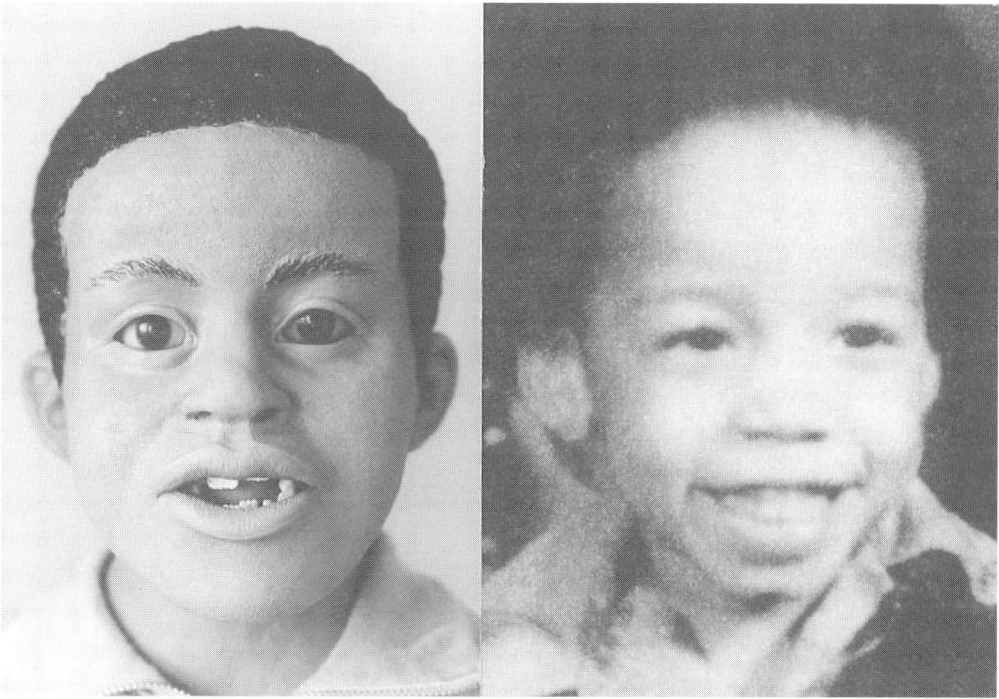

Fig. 8.12 Facial reconstruction (left) of a nine-year-old boy (right) by Gatliff. Courtesy of Karen Taylor (2001).

identification purposes. Previous juvenile forensic investigations have not shed light upon the level of accuracy of the facial reconstruction of children (see Fig. 8.12). Often there is an age difference between the available image of the identified child and the age at death and, even if this is only six months, it may have a large effect upon the appearance of the face, making comparison difficult. The paucity of juvenile tissue depth data has been problematic in the past, so that reconstructions of children have often employed adult tissue depth data (see Fig. 8.13). This has lead to the production of reconstructions lacking the chubby cheeks and jawlines of adolescent faces.

Research at the University of Manchester (Wilkinson & Whittaker, 2002) suggested that juvenile facial reconstruction is as accurate as adult facial reconstruction. Five juvenile reconstructions were produced as part of a single-blind accuracy study. The individuals were aged between eight and 18 years, and the White

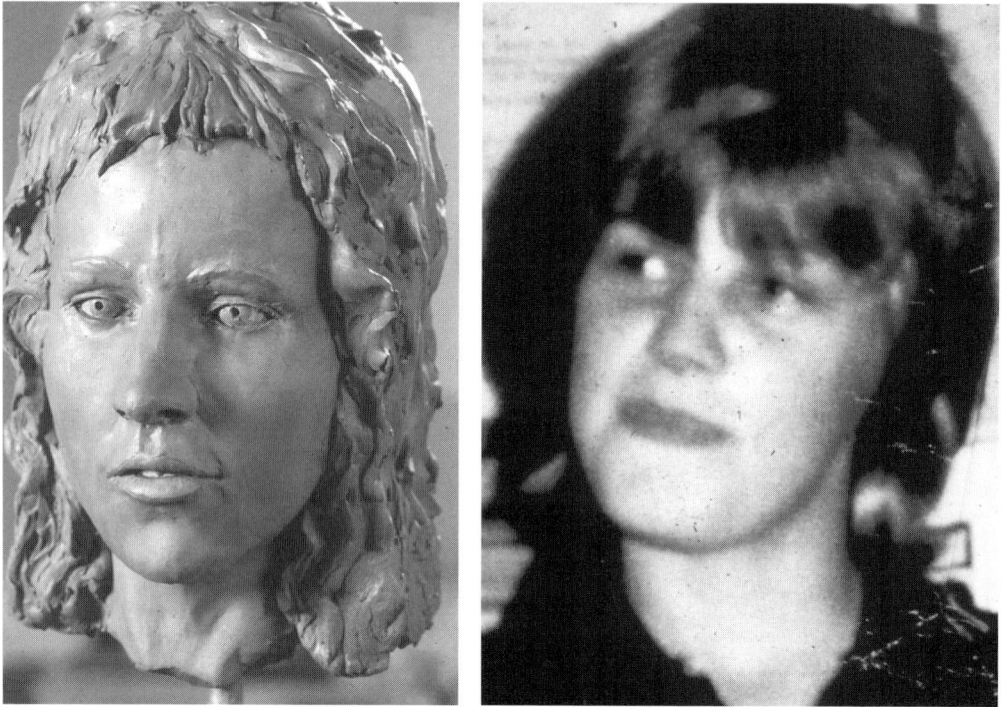

Fig. 8.13 Facial reconstruction of a 15-year-old girl (left) and identified individual (right) by Neave (Prag & Neave, 1997).

British data (Wilkinson, 2002) from 11 to 18-year-olds were employed. A face pool of ten individuals was created using images of the five identified individuals and five females of similar age and ethnic origin. All the photographs were A5 size, black and white images, of the head and shoulders, from frontal or three-quarter views. The five reconstructions were photographed from frontal and three-quarter views using two different lighting techniques. Fifty volunteers were asked to choose a face from the face pool that most resembled each reconstruction. A resemblance assessment was then performed by the same volunteers. The volunteers were shown six comparison sheets, depicting a reconstruction and the identified individual, taken from as similar angles of view as possible (see Fig. 8.14). The volunteers were asked to rate each reconstruction in terms of resemblance to the comparable individual, using the following five-point rating scale: 1 = great, 2 = close, 3 = approximate, 4 = slight, 5 = no resemblance.

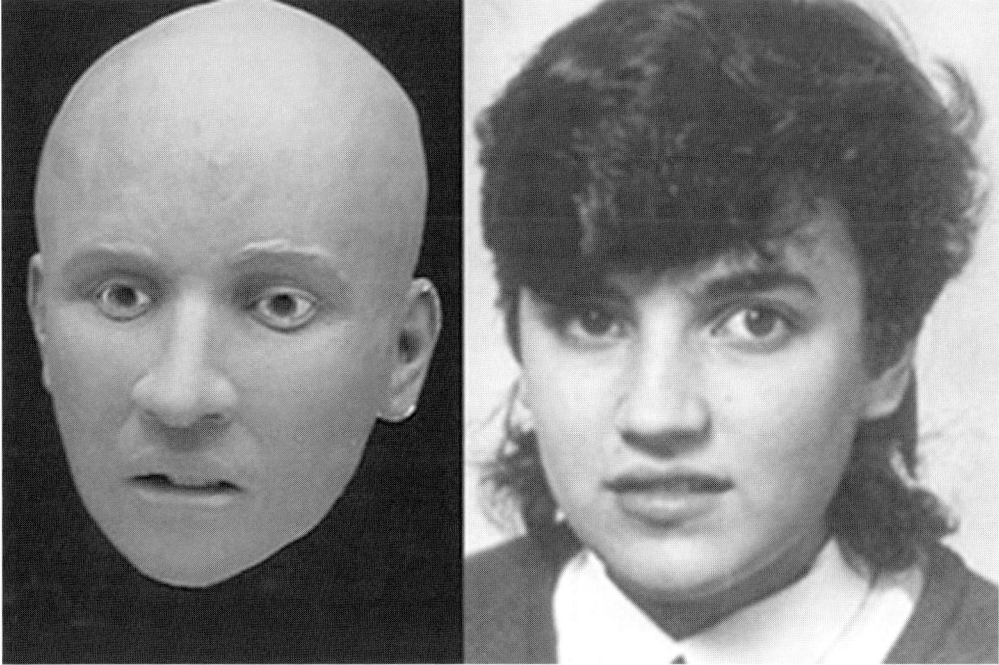

Fig. 8.14 Comparison of a facial reconstruction (left) and identified individual (right) (Wilkinson & Whittaker, 2002).

For each reconstruction the most frequently chosen individual was the correct identification, and these results were significantly different from chance. The overall hit rate was 44%, which was 34% above chance (see Tables 8.11 and Fig. 7.9). Two reconstructions received majority hit rates, and two received hit rates at least 16% greater than any other face. Only a few studies have used face pool identification for the assessment of the accuracy of adult facial reconstruction, with varying degrees of success, and the hit rates of the Manchester study are higher than all the previous studies. Snow *et al.* (1970) produced an average hit rate of 33% above chance; Van Rensburg (1993) produced 19% above chance, and Stephan and Henneberg (2001) recorded 19% below chance. The overall likeness ratings for the reconstructions were 14% great, 42% close, 28% approximate, 14% slight and 2% no resemblance (see Table 8.12 and Fig. 7.7). The five reconstructions were rated as close overall resemblance, with the age and mouth resemblances showing the worst ratings and the body constitution, eyes and nose

Table 8.11 Results of a face pool identification study of juvenile facial reconstruction (FR).

Face Pool	FR1 M n=29	FR1 F n=21	FR1 NE n=19	FR1 E n=31	FR1 Total n=50	FR2 M n=29	FR2 F n=21	FR2 NE n=19	FR2 E n=31	FR2 Total n=50	FR3 M n=29	FR3 F n=21	FR3 NE n=19	FR3 E n=31	FR3 Total n=50	FR4 M n=29	FR4 F n=21	FR4 NE n=19	FR4 E n=31	FR4 Total n=50	FR5 M n=29	FR5 F n=21	FR5 NE n=19	FR5 E n=31	FR5 Total n=50
1	17.5	24.0	23.0	16.0	20.0	7.0	5.0	7.0	5.5	6.0	**45.0**	**33.5**	**29.0**	**63.0**	**40.0**	0.0	0.0	0.0	0.0	0.0	3.5	0.0	0.0	5.0	2.0
2	10.0	4.5	6.5	11.0	8.0	3.5	0.0	0.0	5.5	2.0	14.0	0.0	10.0	5.0	8.0	17.0	5.0	10.0	16.0	12.0	0.0	0.0	0.0	0.0	0.0
3	3.5	4.5	6.5	0.0	4.0	0.0	0.0	0.0	0.0	0.0	17.0	33.5	26.0	16.0	24.0	0.0	5.0	3.0	0.0	2.0	**66.0**	**43.0**	**42.0**	**80.0**	**56.0**
4	10.0	4.5	10.0	5.0	8.0	3.5	0.0	3.0	0.0	2.0	7.0	0.0	3.0	5.0	4.0	17.0	**43.0**	26.0	31.0	28.0	0.0	0.0	0.0	0.0	0.0
5	10.0	4.5	3.0	16.0	8.0	21.0	28.5	32.0	10.5	24.0	10.0	9.5	13.0	5.5	10.0	0.0	9.5	7.0	0.0	4.0	3.5	5.0	6.5	0.0	4.0
6	3.5	0.0	0.0	5.0	2.0	3.5	5.0	7.0	0.0	4.0	0.0	4.5	3.0	0.0	2.0	17.0	19.0	19.0	16.0	18.0	0.0	5.0	3.0	0.0	2.0
7	3.5	0.0	3.0	0.0	2.0	**48.0**	**57.0**	**42.0**	**68.5**	**52.0**	0.0	0.0	0.0	0.0	0.0	0.0	0.0	0.0	0.0	0.0	7.0	28.5	25.5	0.0	16.0
8	**35.0**	**43.0**	**35.5**	**42.0**	**38.0**	3.5	0.0	3.0	0.0	2.0	7.0	9.5	9.5	5.5	8.0	4.0	0.0	3.0	0.0	2.0	3.0	4.5	3.0	5.0	4.0
9	7.0	10.0	9.5	5.0	8.0	3.0	5.0	3.0	5.0	4.0	0.0	0.0	0.0	0.0	0.0	4.0	0.0	0.0	5.0	2.0	10.0	4.5	10.0	5.0	8.0
10	0.0	5.0	3.0	0.0	2.0	7.0	0.0	3.0	5.0	4.0	0.0	9.5	6.5	0.0	4.0	**41.0**	19.0	**32.0**	**32.0**	**32.0**	7.0	9.5	10.0	5.0	8.0
Chi-squared tests ($p < 0.05$)	0.378		0.431			0.370		0.63			0.300		0.010			0.85		0.606			0.96		0.011		

Note: M = male, F = female, NE = non-expert volunteers, E = expert volunteers. Bold = correct identification.

Source: Wilkinson and Whittaker (2002).

Table 8.12 Results of a resemblance rating study for juvenile facial reconstruction (FR).

		Resemblance rating (%)					MODE					Non-parametric tests	
		1-great	2-close	3-approx	4-slight	5-none	Male	Female	Non-expert	Expert	Total	Bet sexes	Bet expertise
FR1	Age	8	**32**	26	22	12	3	2	2	3	2	0.67	0.071
	Body fat	20	**46**	32	2	0	2	1	2	2	2	0.352	0.268
	Eyes	6	**32**	30	26	6	3	2	2	2	2	0.943	0.124
	Nose	12	**28**	26	20	14	2	1	4	2	2	0.872	0.055
	Mouth	6	24	**30**	28	12	3	2	4	3	3	0.111	0.269
	Chin	22	**46**	18	12	2	2	2	2	2	2	0.73	0.686
	Overall	6	**34**	34	22	4	2	3	3	2	2	0.877	0.045
FR2	Age	10	**32**	24	20	14	2	2	2	2	2	0.832	0.984
	Body fat	12	**56**	26	4	2	2	2	2	2	2	0.726	0.548
	Eyes	6	**48**	22	14	10	2	2	2	2	2	0.179	0.435
	Nose	16	**44**	20	16	4	2	2	2	2	2	0.11	0.916
	Mouth	2	18	**42**	24	14	3	3	3	3	3	0.255	0.528
	Chin	14	**52**	18	12	4	2	1	2	2	2	0.107	0.168
	Overall	10	**34**	**36**	16	4	2	3	2	3	3	0.099	0.271
FR3	Age	8	**34**	32	12	14	2	2	2	2	2	0.894	0.09
	Body fat	12	**56**	26	6	0	2	2	2	2	2	0.424	0.548
	Eyes	32	**54**	6	8	0	2	1	2	2	2	0.057	0.534
	Nose	20	**38**	32	8	2	2	2	2	3	2	0.563	0.636
	Mouth	16	**54**	18	8	4	2	2	2	2	2	0.291	0.33
	Chin	34	**40**	20	6	0	2	1	2	1	2	0.95	0.168
	Overall	16	**58**	16	10	0	2	2	2	2	2	0.644	0.358
FR4	Age	18	**44**	20	14	4	2	2	2	2	2	0.604	0.689
	Body fat	18	**48**	28	6	0	2	2	2	2	2	0.68	0.863
	Eyes	20	**42**	30	6	2	2	2	2	1	2	0.473	0.791
	Nose	36	**38**	16	8	2	1	2	2	1	2	0.308	0.267
	Mouth	2	**42**	32	22	2	2	2	2	2	2	0.352	0.323
	Chin	16	**40**	22	18	4	2	2	2	2	2	0.114	0.458
	Overall	18	**44**	22	16	0	2	2	2	2	2	0.633	0.134
FR5	Age	14	**38**	20	16	12	2	2	2	2	2	0.333	0.295
	Body fat	30	**40**	20	10	0	2	1	2	2	2	0.959	0.43
	Eyes	24	**50**	22	2	2	2	2	2	2	2	0.523	0.509
	Nose	18	**62**	14	4	2	2	2	2	2	2	0.768	0.245
	Mouth	0	**42**	34	18	6	2	2	2	3	2	0.698	0.958
	Chin	14	**40**	34	8	4	3	2	2	2	2	0.685	0.619
	Overall	20	**42**	30	4	4	2	2	3	2	2	0.819	0.088

Note: Bold = highest %.
Source: Wilkinson and Whittaker (2002).

resemblances showing the best ratings. A similar adult resemblance assessment was carried out by Helmer *et al.* (1993), where the reconstructions were rated as 38% close, 17% approximate and 42% slight resemblance to the actual individuals. They were rated as a good likeness with regard to age and sex, whilst showing discrepancies at the mouth and eye regions. Even though the adult study had the added advantage of the correct hairstyle and skin tone, the results were less successful than the juvenile study. These results suggest that the British juvenile tissue depth data and the Manchester method of 3-D juvenile facial reconstruction can produce a good resemblance to the individual. Although it is not possible to create a portrait, the reconstruction should be recognisable to someone who knew that person well, such as a family member or close friend.

A recent forensic case, involving the author, illustrates the accuracy of juvenile facial reconstruction and some of the issues relating to this field. In 2001, the Netherlands Forensic Institute approached the author to produce a facial reconstruction of a young girl, whose remains had been found in a lake in The Netherlands. The case was known as the Nulde Girl investigation, and the police had few clues to her identity. The age estimation, from dental assessment, was five-and-a-half years. Prior to this case, there was some degree of reticence regarding the facial reconstruction of children younger than eight years, as the face does not fully develop until that age, and very young children tend to resemble each other. However, an assessment of the skull was carried out, and several characteristic facial details were apparent. The occlusion of the teeth was a Class II, with a significant overjet of the maxillary teeth over the mandibular teeth. The maxillary incisors were prominent and there was a large gap between the two first maxillary incisors. The nasal spine was horizontal, which is unusual in a child of this age, and the eyes were upward-sloping laterally. The skull also showed ambiguous racial origin. Since these details were unusual, the reconstruction went ahead, in the hope that these details would encourage her recognition and identification. White, British, female data were used in the eleven-to-twelve-year age group (Wilkinson, 2002). The reconstruction was produced following the Manchester method, and the Dutch Police presented the head at a major press conference towards the end of the year. After a high-profile forensic campaign, the reconstruction was recognised by two relatives, and the Nulde Girl was eventually

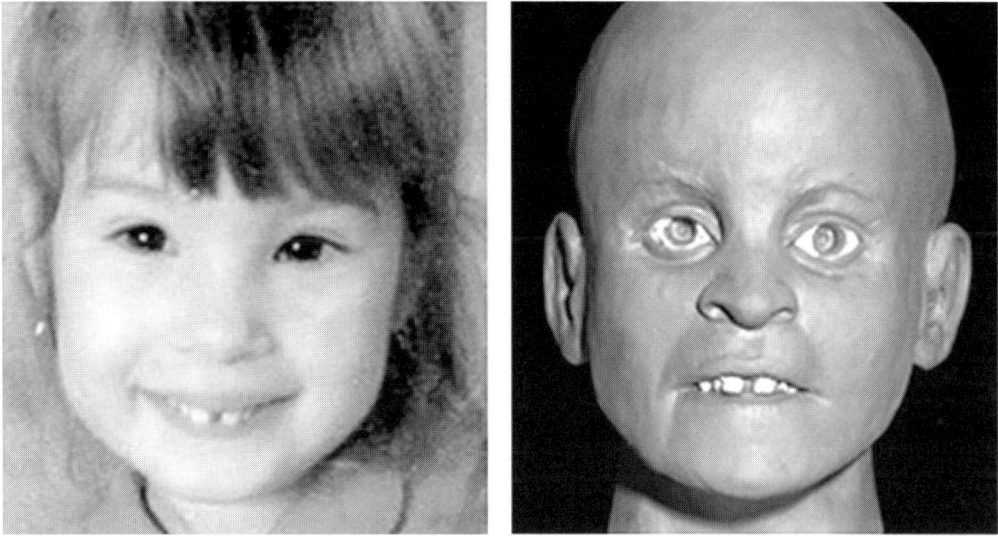

Fig. 8.15 Forensic facial reconstruction of a five-year-old girl (right) and identified individual (left) by Wilkinson.

identified (see Fig. 8.15). The characteristic details of the girl's face are clear in both the reconstruction and the photograph of the girl. This case confirmed that juvenile facial reconstruction is a valuable forensic identification tool, and can be considered as accurate as adult facial reconstruction.

The relationship between the soft and hard tissues of the face in children has not been studied in any detail (other than in the area of orthodontics), also juvenile facial tissue depth data are sparse, and there is very little information regarding juvenile ethnic variation. Further juvenile facial anthropology research would be invaluable to this field.

References

Acsadi, G. and Nemeskeri, J. (1970). *History of Human Lifespan and Mortality*. Budapest: Akademai Kiado.

Altemus, L. A. (1963). Comparative integumental relationships. *Angle Orthodontics* **33**, 3, 217–21.

Anand, R. S., Basiotis, P. B. and Klein, B. W. (1999). Profile of overweight children. *Insight* **13**: http://www.usda.gov/cnpp/Insights/ins13a.PDF.

Angel, J. L. (1978). Restoration of head and face for identification. *Proceedings of Meetings of American Academy of Forensic Science*, St Louis, MO.

Angel, J. L. and Cherry, D. (1977). Personality reconstruction from unidentified remains. *FBI Law Enforcement Bulletin*, 12–15 August.

Arridge, S., Moss, J. P., Linney, A. D. and James, D. R. (1985). Three dimensional digitisation of the face and skull. *Maxillo-Facial Surgery* **13**, 136–43.

Aufderheide, A. C. and Rodriguez-Martin, C. (1998). *The Cambridge Encyclopaedia of Human Paleopathology*. Cambridge: Cambridge University Press.

Auslebrook, W. A. and Van Rensburg, J. H. J. (1982). The significance of race determination in facial reconstruction. Abstract of lecture given at 20th Scientific Congress of International Association of Dental Research, *Journal of Dental Research* **67**, 783.

Auslebrook, W. A., Becker, P. J. and Iscan, M. Y. (1996). Facial soft tissue thicknesses in the adult male Zulu. *Forensic Science International* **79**, 83–102.

Bahrick, H. P., Bahrick, P. O. and Wittlinger, R. P. (1975). Fifty years of memory for names and faces: a cross-sectional approach. *Journal of Experimental Child Psychology* **57**, 377–96.

Baker, M. L. and Dalrymple, G. V. (1978). Biological effects of diagnostic ultrasound: a review. *Radiology* **126**, 479–83.

Bang, G. and Ramm, E. (1970). Determination of age in humans from root dentin transparency. *Acta Odontologiea Scandanavica* **2**, 3–35.

Bankowski, I. M. (1958). Die Bedeutung der Unterkieferform und–stellung für die photographische Schädelidentifizierung. Diss., Frankfurt.

Bartlett, J. C., Hurry, S. and Thorley, W. (1984). Typicality and familiarity of faces. *Memory and Cognition* **12**, 3, 219–28.

BBC News. (1999). UK Fat camp opens its doors. http://news.bbc.co.uk/hi/english/uk/newsid_397000/397323.stm

Behrents, R. (1985). *An Atlas of Growth in the Aging Craniofacial Skeleton.* Craniofacial Growth Series, Ann Arbor: Needham Press.

Bell, C. (1844). *The Anatomy and Philosophy of Expression*, 3rd edn. London: George Bell Publishers.

Berger, D. (1965). *Untersuchungen über die Weichteildickenmaße des Gesichts.* Frankfurt/Main: Diss.

Bertillon, A. (1896). *The Bertillon System of Identification.* In R. W. McClaughry (ed.). Chicago, IL: The Werner Company.

Birkner, F. (1903). Beiträge zur Rassenanatomie der Gesichtsweichteile. *Deutsche Gesellschaft Anthropologie, Ethnologie und Urgeschichte* **34**, 163–5.

　(1905). Beiträge zur Rassenanatomie der Chinesen. *Archiv für Anthropologie*, n.s. IV, 1–40.

　(1907). Die Dicke der Gesichtsweichteile bei verschiedenem Alter, Geschlecht und Rasse. *Sitzungsber Gesellschaft Morphologie, Physiologie München* **23**, 140–6.

Bishara, S. E. (1995). Changes in facial dimensions assessed from lateral and frontal photographs. *American Journal of Orthodontics and Dentofacial Orthopedics* **108**, 489–99.

Bishara, S. E., Jorgensen, G. J. and Jakobsen, J. R. (2000). Facial and dental changes in adolescents and their clinical implications. *The Angle Orthodontist* **70**, 6, 471–83.

Bjork, A. (1947). The face in profile: an anthropometrical x-ray investigation on Swedish children and conscripts. *Svensk Tandlak-T* **49** (Suppl.).

Black, T. K. (1978). Sexual dimorphism in the tooth-crown diameters of the deciduous teeth. *American Journal of Physical Anthropology* **48**, 77–82.

Blair, V. P. (1937). Personal and public reaction to the marred face. *Blue Book of International and Spanish-Speaking Association of Physicians, Dentists and Pharmacists* **8**, 14–17.

Bogren, H. G., Franti, C. E. and Wilmarth, S. S. (1986). Normal variation of the position of the eye in the orbit. *Ophthalmology* **93**, 1072–7.

Booth, R. A. D., Goddard, B. A. and Paton, A. (1966). Measurement of fat thickness in man; a comparison of ultrasound, Harpenden calipers and electrical conductivity. *British Journal of Nutrition* **20**, 719–25.

Borman, H., Ozgur, F. and Gursu, G. (1999). Evaluation of soft tissue morphology of the face in 1050 young adults. *Annals of Plastic Surgery* **42**, 3, 280–8.

Boyd, W. C. (1950). *Genetics and the Races of Man*. Oxford: Blackwell Scientific Publications.

Brennan, S. E. (1985). Caricature generator: the dynamic exaggeration of faces by computer. *Leonardo* **18**, 3, 170–8.

Broadbent, T. R. and Mathews, V. L. (1957). Artistic relationships in surface anatomy of the face. *Plastic and Reconstructive Surgery* **20**, 1, 1–17.

Brothwell, D. R. (1981). *Digging up Bones*. British Museum: Cornell University Press.

  (1989). The relationship of tooth wear to aging. In M. Y. Iscan (ed.), *Age Markers in the Human Skeleton*. Springfield, IL: Charles C. Thomas.

Bruce, V. and Green, P. (1985). *Visual Perception*. London: Lawrence Erlbaum Associates.

Bruce, V. and Valentine, T. (1986). The effect of race, inversion and encoding activity upon face recognition. *Acta Psychologica* **61**, 259–73.

Bruce, V. and Young, A. (1986). Understanding face recognition. *British Journal of Psychology* **77**, 305–27.

Bruce, V. and Young, A. (1998). *In the Eye of the Beholder*. Oxford: Oxford University Press.

Bruce, V., Healey, P., Burton, M. *et al.* (1991). Recognising facial surfaces. *Perception* **20**, 755–69.

Bruce, V., Burton, A. M., Hanna, E. *et al.* (1993). Sex discrimination: how do we tell the difference between male and female faces? *Perception* **22**, 131–52.

Bruce, V., Henderson, Z., Greenwood, K. and Hancock, P. J. B. (1999). Verification of face identities from images captured on video. *Journal of Experimental Psychology* **5**, 339–60.

Brues, A. M. (1958). Identification of skeletal remains. *Journal of Criminal Law, Criminology and Police Science*. **48**, 551–63.

  (1990). The once and future diagnosis of race. In G. W. Gill, and S. Rhine (eds.), *Skeletal Attribution of Race: Methods for Forensic Anthropology*. Maxwell Museum of Anthropology: University of New Mexico, pp. 66–87.

Bullen, B. A., Quaade, F., Olsen, E. and Lund, S. A. (1965). Ultrasonic reflections used for measuring subcutaneous fat in humans. *Human Biology* **37**, 377–84.

Burke, P. H. and Hughes-Lawson, C. A. (1989). Stereophotogrammetric study of growth and development of the nose. *American Journal of Orthodontics and Dentofacial Orthopedics* **96**, 144–51.

Burton, A. M., Bruce, V. and Dench, N. (1993). What's the difference between men and women? Evidence from facial measurement. *Perception* **22**, 153–76.

Caldwell, M. C. (1981). The Relationship of the Details of the Human Face to the Skull and Its Application. M. A. thesis. Arizona State University.

Charney, M. and Coffin, J. C. (1981). Facial reconstruction: a composite procedure. *Proceedings of Meetings of American Academy of Forensic Science*, St Louis MO.

Clement, J. G. and Ranson, D. L. (1998). *Craniofacial Identification in Forensic Medicine*. Sydney: Arnold Publishers.

Clifford, B. R. and Bull, R. (1978). *The Psychology of Person Identification*. London: Routledge and Kegan Paul.

Cobb, M. W. (1955). The age incidence of suture closure. *American Journal of Physical Anthropology* **13**, 394.

Cole, S. (1965). *Races of Man*. London: British Museum Press.

Cole, S. A. (2001). *Suspect Identities: a History of Fingerprinting and Criminal Identification*. London: Harvard University Press.

Colledge, H. (1996). Loss of face? The Effect on the Outcome of Craniofacial Reconstruction when Part of the Skull is Missing. M. Sc. thesis. University of Manchester.

Cook, S. W. (1939). The judgement of intelligence from photographs. *Journal of Abnormal and Social Psychology* **34**, 384–9.

Coon, C. S., Garn, S. M. and Birdsell, J. B. (1950). *Races: A Study of the Problems of Race Formation in Man*. Springfield: C. C. Thomas.

Cooper, J. M. and Hutchinson, D. S. (1997). *Plato: The Complete Works*. Cambridge, IN: Hackett Publishing Company.

Cox, M. and Mays, S. (2000). *Human Osteology in Archaeology and Forensic Science*. London: Greenwich Medical Media.

Czekanowski, J. (1907). Untersuchungen über das Verhaltnis der Kopfmaße zu dem Schädel-maßen. *Archiv für Anthropologie* **34**, 42–89.

Darwin, C. (1872). *The Expression of the Emotions in Man and Animals*. London: John Murray.

Davies, A., Ellis, H. and Shepherd, J. (1981). *Perceiving and Remembering Faces.* London: Academic Press.

Davies, D. M. (1972). *The Influence of Teeth, Diet and Habits on the Human Face.* London: William Heinemann Medical Books Ltd.

Dean, M. C. and Beynon, A. D. (1991). Histological reconstruction of crown formation times and initial root formation times in the modern human child. *American Journal of Physical Anthropology* **86**, 215–88.

della Porta, G. (1586). *De humana physiognomonia.*

Diedrich, F. (1926). Ein Beitrag zur Prüfung der Leistungsfähigkeit der plastischen Rekonstruktionsmethode der Physiognomie bei der Identifizierung von Schädeln. *Deutsche Zeitschrift fur Gesellschaft Gerichtliche Medizin* **8**, 365–89.

Dibbets, J. M. H. and Nolte, K. (2001). Comparison of linear cephalometric dimensions in Americans of European descent (Ann Arbor, Cleveland, Philadelphia) and Americans of African descent (Nashville). *Angle Orthodontist* **72**, 4, 324–30.

Dion, K., Berscheid, E. and Walster, E. (1972). What is beautiful is good. *Journal of Personality and Social Psychology* **24**, 285–90.

Ditch, L. E. and Rose, J. C. (1972). A multivariate dental sexing technique. *American Journal of Physical Anthropology* **37**, 61–4.

Dumont, E. R. (1986). Midfacial tissue depths of white children: an aid in facial feature reconstruction. *Journal of Forensic Science* **31**, 4, 1463–9.

Dunn, K. W. and Harrison, R. K. (1997). Naming of parts: a presentation of facial surface anatomical terms. *British Journal of Plastic Surgery*, **50**, 584–9.

Dunn, L. C. (1967). Race and biology. In N. Korn and F. Thompson (eds.), *Human Evolution.* New York: Holt, Reinhert & Winston.

Ekman, P. and Friesen, W. V. (1971). Constants across cultures in the face and emotion. *Journal of Personality and Social Psychology* **17**, 2, 124–90.

Ellis, H. D., Deregowski, I. B. and Shepherd, J. W. (1975). Descriptions of white and black faces by white and black subjects. *International Journal of Psychology* **10**, 2, 119–23.

Ellis, H. D., Shepherd, J. W. and Davies, G. M. (1979). Identification of familiar and unfamiliar faces from internal and external features: some implications for theories of face recognition. *Perception* **8**, 431–9.

El-Nofely, A. (1972). Anthropometric study of growth changes of some head and face measurements in an Egyptian group. *Egyptian Dental Journal* **18**, 2, 141–50.

Enlow, D. H. (1982). *Handbook of Facial Growth*, 2nd edn. Philadelphia, PA: W. B. Saunders.

Enlow D. H. and Hans M. G. (1996). *Essentials of Facial Growth*. Philadelphia, PA: W. B. Saunders.

Evenhouse, R. M., Rasmussen, M. and Sadler, L. (1992). Computer-aided forensic facial reconstruction. *Journal of Biological Chemistry* **19**, 22–8.

Evison, M. P. (2002). Torticollis in an unidentified female from Leeds, England. *Proceedings of the 10th Conference of the International Association of Craniofacial Identification*, Bari, Italy.

Evison, M. P., Finegan, O. M. and Blythe, T. C. (1999). Computerised 3-D facial reconstruction: research update. *Assemblage* **4**: http://www.shef.ac.uk/assem/4/evison.html

Fagan, J. F. (1972). Infants' recognition memory for faces. *Journal of Experimental Psychology* **14**, 453–76.

Farkas, L. G. (1981). *Anthropometry of the Head and Face in Medicine*. New York: Elsevier.

(1994). Asymmetry of the head and face. In *Anthropometry of the Head and Face*, 2nd edn. New York, NY: Raven Press, pp. 103–11.

Farkas, L. G and Hreczko, T. A. (1994). Age-related changes in selected linear and angular measurements of the craniofacial complex in healthy North American Caucasians. In L. G. Farkas (ed.), *Anthropometry of the Head and Face*, 2nd edn. New York, NY: Raven Press, pp. 89–102.

Farkas, L. G. and Posnick, J. C. (1992). Growth and development of the head. *Cleft-Palate Craniofacial Journal* **29**, 301–29.

Farkas, L. G., Hreczko, T. A., Kolar, J. C. and Munro, I. R. (1985a). Vertical and horizontal proportions of the face in young adult North American Caucasians: revision of neoclassical canons. *Plastic and Reconstructive Surgery* (March) **75**, 3, 328–37.

Farkas, L. G., Sohm, P., Kolar, J. C., Katic, M. J. and Munro, I. R. (1985b). Inclinations of the facial profile: art versus reality. *Plastic and Reconstructive Surgery*, **75**, 4, 509–19.

Farkas, L. G., Forrest, C. R. and Litsas, L. (2000). Revision of neoclassical facial canons in young adult Afro-Americans. *Aesthetic Plastic Surgery* **24**, 3, 179–84.

Fedosyutkin, B. A. and Nainys, J. V. (1993). The relationship of skull morphology to facial features. In Iscan, M. Y. and Helmer, R. P. (eds.), *Forensic Analysis of the Skull*. New York: Wiley-Liss Inc., pp. 199–213.

Feik, S. A. and Glover, J. E. (1998). Growth of children's faces. In Clement and Ranson, 1998.

Ferrario, V. F., Sforza, C. and Serrao, G. (2000). A three-dimensional quantitative analysis of lips in normal young adults. *Cleft-Palate Craniofacial Journal* **37**, 1, 48–54.

Fiorato, V., Boylston, A. and Knusel, C. (2000). *Blood Red Roses: The Archaeology of a Mass Grave from the Battle of Towton, AD 1461*. Oxford: Oxbow Books.

Fischer, E. (1903). Anatomische Untersuchungen an den Kopfweichteilen zweier Papua. *Deutsche Gesellschaft für Anthropologie, Ethnologie und Urgeschichte* **36**, 118–22.

Fitzgerald, C. M. (1999). Do dental microstructures have a regular time dependency? *Journal of Human Evolution* **35**, 371–86.

Ford, C. S., Protho, E. T. and Child, I. L. (1966). Some transcultural comparisons of aesthetic judgement. *Journal of Social Psychology* **68**, 19–25.

Funte & Waynalts. Encyclopedia (1961). New York and London.

Furuta M. (2001). Measurement of orbital volume by computed tomography – especially on the growth of the orbit. *Japanese Journal of Ophthalmology* **45**, 6, 600–6.

Galen, 1956. *Galen on Anatomical Procedure* [De anatomicis administrationibus], trans. C. Singer. Oxford.

Garlie, T. N. and Saunders, S. R. (1999). Midline facial tissue thicknesses of subadults from a longitudinal radiographic study. *Journal of Forensic Science* **44**, 1, 61–7.

Garn, S. M., Lewis, A. B. and Kerewsky, R. S. (1964). Sex difference in tooth size. *Journal of Dental Research* **43**, 306.

Gatliff, B. P. (1984). Facial sculpture on the skull for identification. *American Journal of Forensic Medicine and Pathology* **5**, 4, 327–32.

Gatliff, B. P. and Snow, C. C. (1979). From skull to visage. *Journal of Biocommunication* **6**, 2, 27–30.

Genecov, J. S., Sinclair, P. M. and Dechow, P. C. (1990). Development of the nose and soft tissue profile. *Angle Orthodontist* **60**, 191–8.

George, R. M. (1987). The lateral craniographic method of facial reconstruction. *Journal of Forensic Sciences* **32**, 5, 1305–30.

(1993). Anatomical and artistic guidelines for forensic facial reconstruction. In *Forensic Analysis of the Skull*, Wiley-Liss Inc., pp. 215–27.

Gerasimov, M. M. (1971). *The Face Finder*. New York: Hutchinson.

(1975). *The Reconstruction of the Face from the Basic Structure of the Skull*, trans. W. Tshernezky, Russia: Publishers unknown.

Giles, E. and Elliot, O. (1962). Race identification from cranial measurements. *Journal of Forensic Sciences* **7**, 147–57.

Glaister, J. and Brash, J. C. (1937). *Medico-legal Aspects of the Ruxton Case*. Edinburgh: Elsevier Ltd.

Glanville, E. V. (1969). Nasal shape, prognathism and adaption in man. *American Journal of Physical Anthropology* **30**, 29–38.

Goldhamer, K. (1926). Aus dem Rontgenlaboratorium der I. anatomischen Lehrkanzel. *Anatome Entwiche Desch* **81**.

Gonzalez-Figueroa, A. (1996). Evaluation of the Optical Laser Scanning System for Facial Identification. Ph.D. thesis. University of Glasgow.

Gould, S. J. (1981). *The Mismeasure of Man*. New York: W. W Norton & Co.

Gray, H. (1973). *Gray's Anatomy*, 35th edn. Ed. R. Warwick and P. L. Williams. London: Longman Group Ltd.

(1980). *Gray's Anatomy*, 36th edn. Ed. P. L. Williams and R. Warwick. London: Churchill Livingstone.

Greyling, I. H. and Meiring, J. H. (1993). Morphological study on the convergence of the facial muscles at the angle of the mouth. *Acta Anatomica* **143**, 127–9.

Gustafson, G. (1950). Age determination on teeth. *Journal of the American Dental Association* **41**, 45–54.

Haglund, W. D. and Reay, D. T. (1991). Use of facial approximation in identification of Green River Serial Murder Victims. *American Journal of Forensic Medicine and Pathology* **12**, 2, 132–42.

Hajnis, K., Farkas, L. G., Ngim, R. C. K., Lee, S. T. and Venkatadri, G. (1994). Racial and ethnic morphometric differences in the craniofacial complex. In Farkas, 1994.

Hanihara, T. (2000). Frontal and facial flatness of major human populations. *American Journal of Physical Anthropology* **111**, 1, 105–34.

Harmon, L. (1973). The recognition of faces. *Scientific American* **227** November, 71–82.

Heglar, R. and Parks, C. R. (1980). Juvenile facial restoration. Pediatric and cephalometric expectations. *Proceedings of the American Association of Forensic Science*, Annual meeting, New Orleans.

Helmer, R. (1984). *Schädelidentifizierung durch elektronische Bildmischung.* Heidelberg, Kriminalistik-Verlag.

Helmer, R., Rohricht, S., Petersen, D. and Moer, F. (1989). Plastische Gesichtsrekonstruktion als Möglichkeit der Identifizierung unbekannter Schädel (II) *Archives Kriminology* **184**, 5–6, 142–60.

  (1993). Assessment of the reliability of facial reconstruction. In Iscan and Helmer, 1993, pp. 229–47.

Hill, B., Macleod, I. and Crothers, A. (1996). Rebuilding the face of George Buchanan. *Journal of Audiovisual Media in Medicine* **19**, 1, 11–15.

Hill, B., Macleod, I. and Watson, L. (1993). Facial reconstruction of a 3500-year-old Egyptian mummy using axial computed tomography. *Journal of Audiovisual Media in Medicine* **16**, 11–13.

Hillson, S. (1996). *Dental Anthropology.* Cambridge: Cambridge University Press.

His, W. (1895). Anatomische Forschungen über Johann Sebastian Bach Gebeine und Antlitz nebst Bemerkungen über dessen Bilder. *Abhandlungen der mathematisch-physikalischen Klasse der Königlichen Sachsischen Gesellschaft der Wissenschaften* **22**, 379–420.

Hjalgrim, H., Lynnerup, N., Liversage, M. and Rosenklint, A. (1995). Stereolithography: potential applications in anthropological studies. *American Journal of Physical Anthropology* **97**, 329–33.

Hodson, G., Lieberman, L. S. and Wright, P. (1985). In vivo measurements of facial tissue thicknesses in American Caucasoid children. *Journal of Forensic Science* **30**, 4, 1100–12.

Hogarth, B. (1965). *Drawing the Human Head.* New York: Watson Guptill Publications.

Hogarth, W. (1753). *The Analysis of Beauty.* Oxford: Oxford University Press.

Hooton, E. A. (1946). *Up from the Apes.* New York: Macmillan.

Houghton, P. (1978). Polynesian mandibles. *Journal of Anatomy* **127**, 251–60.

Howells, W. W. (1970). Multivariate analysis for the identification of race from the crania. In T. D. Stewart (ed.), *Personal Identification in Mass Disasters.* Philadelphia, PA: American Museum of Natural History.

Hunt, E. E. and Gleiser, I. (1955). The estimation of age and sex of preadolescent children from bones and teeth. *American Journal of Physical Anthropology* **13**, 479–87.

Hwang, H., Kim, W. and McNamara Jr, J. A. (2001). Ethnic differences in soft tissue profiles of Korean and European-American adults with normal occlusions and well-balanced faces. *Angle Orthodontist* **72**, 1, 72–80.

Iliffe, A. H. (1960). A study of preferences in feminine beauty. *British Journal of Psychology* **51**, 267.

Imai, K. and Tajima, S. (1993). Measurement of normal eyeball position and its application for evaluation of exophthalmos in craniofacial synostosis. *Plastic and Reconstructive Surgery* **92**, 588–92.

Inoue, K., Ichikawa, R., Nagashima, M. and Kodama, G. (1995). Sex differences in the shapes of several parts of the young Japanese face. *Applied Human Science* **14**, 4, 191–94.

Iscan, M. Y. and Helmer, R. P. (1993). *Forensic Analysis of the Skull*. New York, NY: Wiley-Liss Inc.

Jankowsky, W. (1930). Über Unterkiefermasse und ihren rassendiagnostischen Wert. *Zeitschrift für morphologie und anthropogie* **28**, 347–59.

Jantz, R. L., Hunt, D. R., Falsetti, A. B. and Key, P. J. (1992). Variation among North American Indians: analysis of Boas's anthropometric data. *Human Biology* **64**, 435–61.

Jenkinson, J. (1997) Face facts: a history of physiognomy from ancient Mesopotamia to the end of the nineteenth century. *Journal of Biomedical Communication* **24**, 3, 2–7.

Johnson, D. R., O'Higgins, P., Moore, W. J. and McAndrew, T. J. (1990). Determination of race and sex of human skulls by discriminant function analysis of linear and angular dimensions: an appendix. *Forensic Science International* **45**, 1–3.

Kahle, W., Leonhardt, H. and Platzer, W. (1992). *Colour Atlas/Text of Human Anatomy, Vol. 1: Locomotor System*. New York: Thieme Medical Publishers.

Kemp, R., Towel, N. and Pike, G. (1997). Recognising own and other race faces from video surveillance footage. Paper at 7th European Conference of Psychology and Law, Stockholm.

Kilian, J. and Vlček, E. (1989). Age determination from teeth in adult individuals. In M. Y. Iscan (ed.), *Age Markers in the Human Skeleton*. Springfield, IL: Charles C. Thomas, pp. 255–75.

Knight, B. and Whittaker, D. K. (1997). Medical and dental investigations in the Rosemary West case. *Medicolegal Journal* **65**, 107–21.

Kollman, J. (1898). Die Weichteile des Gesichts und die Persistenz der Rassen. *Anatomischer Anzeiger* **15**, 165–77.

Kollman, J. and Buchly, W. (1898). Die Persistenz der Rassen und die Reconstruction der Physiognomie prähistorischer Schädel. *Archiv für Anthropologie*, **25**, 329–59.

Krogman, W. M. and Iscan, M. Y. (1962). *The Human Skeleton in Forensic Medicine*, 1st edn. Springfield, IL: C. C. Thomas Publishers.

Krogman, W. M. and Iscan, M. Y. (1986). *The Human Skeleton in Forensic Medicine*, 2nd edn. Springfield, IL: C. C. Thomas Publishers.

Lanarch, S. L. (1978). Australian aboriginal craniology. *Oceania Monographs* **21**, 1.

Landau, T. (1989). *About Faces*. New York: Anchor Books.

Langlois, J. H., Ritter, J. M., Roggman, L. A. and Vaughn, L. S. (1991). Facial diversity and infant preferences for attractive faces. *Developmental Psychology* **27**, 79–84.

Langlois, J. H., Roggman, L. A. and Musselman, L. (1994). What is average and what is not average in attractive faces? *Psychological Science* **5**, 214–20.

Larsen, C. S. (1997). *Bioarchaeology: interpreting human behaviour from the human skeleton.* Cambridge: Cambridge University Press.

Latta, G. H. (1988). The midline and its relation to anatomic landmarks in the edentulous patient. *Journal of Prosthetic Dentistry* **59**, 681–3.

Latta, G. H., Weaver, J. R. and Conkin, J. E. (1991). The relationship between the width of the mouth, interalar width, bizygomatic width and interpupillary distance in edentulous patients. *Journal of Prosthetic Dentistry* **65**, 250–4.

Lauprecht, E., Scheper, J. and Schroder, J. (1957). Messungen der Speckdicke lebender Schweine nach dem Scholotverfahren. *Mittel Deutsche Landwirtschaftliche-Gesellschaft* **72**, 881.

Lavater J. C. and Holcroft T. (1789). *Essays on Physiognomy*. London: G.G. J. & J. Robinson.

Lebedinskaya, G. U., Balueva, T. S. and Veselovskaya, E. B. (1993). Development of methodological principles for reconstruction of the face on the basis of skull material. In M. Y. Iscan and R. P. Helmer (eds.), *Forensic Analysis of the Skull*. New York: Wiley-Liss Inc., pp. 183–98.

Leopold, D. (1968). *Identifikation durch Schädeluntersuchung unter besonderer Berücksichtigung der Superprojektion.* Leipzig: Habilitationsschrift.

Liversidge H.M., Herdeg B. and Rosing F.W. (1998). Dental age estimates in non-adults. In K.W. Alt, F.W. Rosing and M. Teschler-Nicola (eds.), *Dental Anthropology, Fundamentals, Limits and Prospects*. Vienna: Springer, pp. 419–42.

Lovejoy, C.O. (1985). Dental wear in the Libben population. *American Journal of Physical Anthropology* **68**, 47–56.

Lysell L., Magnusson, B. and Thilander, B. (1962). Time and order of eruption of primary teeth: a longitudinal study. *Odontologisk Revy* **13**, 217–34.

McClintock Robinson, J., Rinchose, D.J. and Zullo, T.G. (1986). Relationship of skeletal pattern and nasal form. *American Journal of Orthodontics* **89**, 499–506.

McGregor, J.H. (1926). Restoring Neanderthal man. *National History* **26**, 288–93.

McKelvie, S.J. (1978). Sex differences in facial memory. In M.M. Gruneburg, P.E. Morris and R.N. Sykes (eds.), *Practical Aspects of Memory*. London: Academic Press, pp. 263–9.

Macho, G.A. (1986). An appraisal of plastic reconstruction of the external nose. *Journal of Forensic Science* **31**, 4, 1391–1403.

Malone, D.R., Morris, H.H., Kay, M.C. and Levin, H.S. (1982). Prosopagnosia: a double dissociation between the recognition of familiar and unfamiliar faces. *Journal of Neurology, Neurosurgery and Psychiatry* **45**, 820–2.

Manhein, M.H., Barsley, R.E., Listi, G.A., Musselman, R., Barrow, N.E. and Ubelaker, D.H. (2000). In vivo facial tissue depth measurements for children and adults. *Journal of Forensic Science* **45**, 1, 48–60.

Mann, R.W., Jantz, R.L., Bass, W.M. and Willey, P.S. (1991). Maxillary suture obliteration. *Journal of Forensic Science* **36**, 781–91.

Martin, J.G. (1964). Racial ethnocentrism and judgement of beauty. *Journal of Social Psychology* **63**, 59.

Mays, S. (1996). The human skeletal remains. In J.R. Timby (ed.), *The Anglosaxon Cemetery at Empingham II, Rutland*. Monograph 70, Oxford: Oxbow, pp. 21–33.

Mays, S. and Cox, M. (2000). Sex determination in skeletal remains. In M. Cox and S. Mays, 2000.

Meredith, H.V. (1973). Gingival emergence of human deciduous teeth; a synoptic report. *Journal of Tropical Pediatrics and Environmental Child Health*. Special issue, 195–9.

Michael, S.D. and Chen, M. (1996). The 3-D reconstruction of facial features using volume distortion. Proceedings of 14th Annual Conference of Eurographics UK, 297–305.

Miles, A.E.W. (1963). The dentition in the assessment of individual age in skeletal material. In D.R. Brothwell (1963). *Digging Up Bones*. British Museum: Cornell University Press, 191–209.

Milgrim, L.M., Lawson, W. and Cohen, A.F. (1996). Anthropometrical analysis of the female Latino nose: revised aesthetic concepts and their surgical implications. *Archives of Otolaryngology: Head and Neck Surgery* **122**, 10, 109–1086.

Molleson, T., Cruse, K. and Mays, S. (1998). Some sexually dimorphic features of the human juvenile skull and their value in sex determination in immature skeletal remains. *Journal of Archaeological Sciences* **25**, 719–28.

Morant, G.M. (1936). A biometric study of the human mandible. *Biometrika* **28**, 84–122.

Moss, J.P., Linney, A.D., Grindrod, S.R., Arridge, S.R. and Clifton, J.S. (1987). Three-dimensional visualisation of the face and skull using computerised tomography and laser scanning techniques. *European Journal of Orthodontics* **9**, 247–53.

Neave, R.A.H. (1979). Reconstruction of the heads of three Egyptian mummies. *Journal of Audiovisual media in Medicine* **2**, 156–64.

(1994). Book review – Forensic analysis of the skull. *International Journal of Osteoarchaeology* **4**, 163.

(1998). Age changes in the face in adulthood. In J.G. Clement and D.L. Ranson (eds.), *Craniofacial Identification in Forensic Medicine*, Sydney: Arnold Publications, pp. 215–231.

Nelson, L.A. and Michael, S.D. (1998). The application of volume deformation to 3-D facial reconstruction; a comparison with previous techniques. *Forensic Science International* **94**, 167–81.

Nute, S.J. and Moss, J.P. (2000). Three-dimensional facial growth studied by optical surface scanning. *Journal of Orthodontics* **27**, 1, 31–38.

Ofodile, F.A., Bokhari, F.J. and Ellis, C. (1993). The Black American nose. *Annals of Plastic Surgery* **31**, 3, 209–18.

Parsons, F.G. and Boc, L.R. (1905). The relation of cranial sutures to age. *Journal of the Royal Anthropology Institute* **25**, 30–38.

Pearson, K. (1926). The skull and portraits of George Buchanan. *Forensic Science International* **83**, 51–9.

Pearson, K. (1928). The skull and portraits of Henry Stewart, Lord Darnley, and their bearing on relationship to busts. *Biometrika* **20**, 1–14.

Peck, H. and Peck, S. (1970). A concept of facial aesthetics. *Angle Orthodontist* **40**, 4, 284–318.

Pedersen, P. O. (1949). The East Greenland Eskimo Dentition. Kobenhavn, 1940. *Bianco Lunos, Banstrijkken and Meddelsen on Gronland* **60**, 142–244.

Penry, J. (1939). *How to Judge Character from the Face.* London: Hutchinson. (1971). *Looking at Faces.* London: Elek Books.

Phillips, V. M. and Smuts, N. A. (1996). Facial reconstruction; utilisation of computerised tomography to measure facial tissue thickness in a mixed population. *Forensic Science International* **83**, 51–9.

Porter, J. P. and Olsen, K. L. (2001). Anthropometric facial analysis of the African-American woman. *Archives of Facial Plastic Surgery* **3**, 3, 191–7.

Potclays Ltd (2003). Suppliers of ceramic pot clay. Stoke on Trent.

Pounder, D. J. (1984). Forensic aspects of aboriginal skeletal remains in Australia. *American Journal of Forensic Medicine and Pathology* **5**, 1, 41–52.

Prag, J. and Neave, R. A. H. (1997). *Making Faces.* London: British Museum Press.

Preeyanont, P. (1995). The standard angle between the longitudinal axis of the ear and the bridge of the nose in Thai women. *Journal of the Medical Association of Thailand* **78**, 3, 127–34.

Prokopec, M. and Ubelaker, D. H. (2002). Reconstructing the shape of the nose according to the skull. *Forensic Science Communications* **4**, 1.

Ramsey, N., Bull, R. and Gahagan, D. (1982). The effects of facial disfigurement on the proxemic behaviour of the general public. *Journal of Applied Social Psychology* **12**, 137–50.

Rathbun, T. A. (1984). *Personal Identification: Facial Reproduction in Human Identification Case Studies in Forensic Anthropology.* Springfield, IL: C. C. Thomas, pp. 343–56.

Redfield, A. (1970). A new aid to aging immature skeletons; development of the occipital bone. *American Journal of Physical Anthropology* **33**, 207–20.

Rhine, J. S. and Campbell, H. R. (1980). Thickness of facial tissues in American Blacks. *Journal of Forensic Science* **25**, 4, 847–58.

Rhine, J. S., Moore, C. E. and Westin, J. T. (eds.) (1982). *Facial Reproduction: Tables of Facial Tissue Thickness of American Caucasoids in Forensic Anthropology*. Maxwell Museum, Technical Series, no. 1, University of New Mexico.

Rhine, S. (1983). Tissue thickness for South-western Indians. Ph.D. thesis. Physical Anthropology Laboratories. Maxwell Museum, University of New Mexico.

Riola, M. L., Moyers, R. E., Macnamara, J. A. and Hunter, W. S. (1974). *An Atlas of Craniofacial Growth. Monograph 2, Craniofacial Growth Series*. Ann Arbor, Center for Human Growth and Development, University of Michigan.

Rushton, M. A. (1933). On the fine contour lines of the enamel of milk teeth. *Dental Record* **53**, 170–1.

Sahni, D. (2002). Preliminary study on facial soft tissue thickness by magnetic resonance imaging in Northwest Indians. *Forensic Science Communications* **4**, 1.

Samuels, C. A., Butterworth, G., Roberts, T., Grauper, L. and Hole, G. (1994). Facial aesthetics: babies prefer attractiveness to symmetry. *Perception* **23**, 823–31.

Sanchez, A. E. (1980). Rhinoplasty on the mestizo nose. *Aesthetic Plastic Surgery* **77**, 239–52.

Sauer, N. J. (1992). Forensic Anthropology and the concept of race – if races don't exist, why are forensic anthropologists so good at identifying them? *Social Science and Medicine* **34**, 2, 107–11.

Schnalke, T. (1995). *Diseases in Wax – The History of the Medical Moulage*. Berlin: Quintessence Publishing Co., Inc.

Schour I. (1936). The neonatal line in the enamel and dentin of the human deciduous teeth and first premolar. *Journal of the American Dental Association* **23**, 1946–55.

Schour, I. and Massler, M. (1944). *Development of Human Dentition*. Chicago University of Illinois School of Dentistry.

Schultz, A. H. (1918). Relation of the external nose to the bony nose and nasal cartilages in whites and negroes. *American Journal of Physical Anthropology* **1**, 3, 329–38.

Schutowski, H. (1993). Sex determination of infant and juvenile skeletons: I morphognostic features. *American Journal of Physical Anthropology* **90**, 199–205.

Scott, P. (1998). An investigation into the relationship between hard and soft tissues in the lip–chin region of the face. Anatomical Sciences. B.Sc. thesis, University of Manchester.

Secord, P. F., Bevan, W. and Katz, B. (1956). The negro stereotype and perceptual accentuation. *Journal of Abnormal and Social Psychology* **53**, 78–83.

Shepherd, J. (1981). Social factors in face recognition. In G. Davies, H. Ellis and J. Shepherd (eds.), *Perceiving and Remembering Faces.* London: Academic Press.

Sim, R. S. T. and Smith, J. D. (2000). Comparison of the aesthetic facial proportions of Southern Chinese and White Women. *Archives of Facial Plastic Surgery* **2**, 113–20.

Simpson, E. and Henneberg, M. (2002). Variation in soft tissue thicknesses on the human face and their relationship to craniometric dimensions. *American Journal of Physical Anthropology* **118**, 121–33.

Skiles, M. S. and Randall, P. (1983). The aesthetics of ear placement. *Plastic and Reconstructive Surgery* **84**, 8, August, 133–8.

Slater, A., Bremner, G., Johnson, S. P., Sherwood, P., Hayes, R. and Brown, E. (2000). Newborn infants' preference for attractive faces. *Infancy* **1**, 265–74.

Smith, S. L. and Buschang, P. H. (2001). Midsaggital facial tissue thicknesses of children and adolescents from the Montreal Growth Study. *Journal of Forensic Science* **46**, 6, 1294–302.

Smith, W. S. (1961). *Ancient Egypt.* Boston, MA: Beacon Press.

Snow C. E. (1974). *Early Hawaiians.* Lexington, KY: University of Kentucky Press.

Snow, C. C., Gatliff, B. P. and McWilliams, K. R. (1970). Reconstruction of facial features from the skull: an evaluation of its usefulness in forensic anthropology. *American Journal of Physical Anthropology* **33**, 221–8.

Sobotta, J. (1983). *Atlas of Human Anatomy 1: Head, Neck and Upper Extremities*, 10th English edn. Ed. H. Ferner and J. Staubesand. Baltimore-Munich: Urban and Schwarzenberg.

Spencer, M. A. and Demes, B. (1993). Biomechanical analysis of masticatory system configuration in Neanderthals and Inuits. *American Journal of Physical Anthropology* **91**, 1–20.

Stadtmuller, F. (1922). Zur Beurteilung der plastischen Rekonstruktionsmethode der Physiognomie auf dem Schädel. *Zeitschrift für Morphologie und Anthropologie* **22**, 337–72.

(1923). Plastische Physiognomie-Rekonstruktionen auf den beiden diluvialen Schädeln von Obercassel bei Bonn. *Zeitschrift für Morphologie und Anthropologie* **23**, 301–14.

Stenstrom S. (1946). Untersuchungen über die Variation und Kovariation der optischen Elemente des menschlichen Auges. *Acta Ophthalmologica Scandinavia Supplementum* **26**, 1–103.

Stephan, C. N. (2000). Do resemblance ratings measure a facial approximation accuracy? *Proceedings of 9th Biennial Conference of the International Association of Craniofacial Identification*, Washington.

(2002). Facial approximation: globe projection guideline falsified by exophthalmometry literature. *Journal of Forensic Science* **47**, 4, 730–5.

Stephan, C. and Henneberg, M. (2001). Building faces from dry skulls: are they recognised above chance rates? *Journal of Forensic Science* **46**, 3, 432–40.

Stewart, T. D. (1948). Medicolegal aspects of the skeleton – age, sex, race and stature. *American Journal of Physical Anthropology* **6**, 315–21.

(1979). *Essentials of Forensic Anthropology*. Springfield, IL: C. C. Thomas Publications.

(1983). The points of attachment of the palpebral ligaments; their use in facial reconstructions on the skull. *Journal of Forensic Science* **28**, 4, 858–63.

Stouffer, J. R. (1963). The relationship of ultrasonic measurements and x-rays to body composition. *Annals of the New York Academy of Sciences* **110**, 31–9.

Suk, V. (1935). Fallacies of anthropological identifications and reconstructions: a critique based on anatomical dissections. *Publications of the Faculty of Science, University of Masaryk, Brno*, 207, pp. 1–18.

Sutton, P. R. N. (1969). Zygomatic diameter: the thickness of the tissues over the zygions. *American Journal of Physical Anthropology* **30**, 303–10.

Suzuki, K. (1948). On the thickness of the soft parts of the Japanese face. *Journal of Anthropology of the Society of the Nippon* **60**, 7–11.

Tandler, J. (1909). Über den Schädel Haydns. *Mitteilungen der Anthropologie Gesellschaft Wien* **39**, 260–80.

Tanner, J. M. (1952). The assessment of growth and development in children. *Archives of Diseases in Childhood* **27**, 10–33.

Taylor, J. A. and Angel, C. (1998). Facial reconstruction and approximation. In J. G. Clement and D. L. Ranson (eds.), *Craniofacial Identification in Forensic Medicine*. Sydney: Arnold Publishers.

Taylor, K. (2001). *Forensic Art and Illustration*. Boca Raton: CRC Press.

Tian S., Yasuhiro N., Isberg B. and Lennerstrand G. (2000). MRI measurements of normal extraocular muscles and other orbital structures. *Graefe's Archives of Clinical and Experimental Ophthalmology* **238**, 393–404.

Tiranti, Alec – Suppliers of sculptural materials. The High Street, Reading.

Todd, J. T., Mark, L. S., Shaw, R. E. and Pittenger, J. B. (1980). The perception of human growth. *Scientific American* **242**, 2, 132–44.

Todd, T. W. and Lindala, A. (1928). Thickness of the subcutaneous tissues in the living and the dead. *American Journal of Anatomy* **41**, 2, 153–95.

Todd, T. W. and Lyon, D. W. Jr (1924). Endocranial suture closure, its progress and age relationship: Part I. Adult males of White stock. *American Journal of Physical Anthropology* **7**, 325–84.

  (1925). Cranial suture closure. Part II: Ectocranial closure in adult males of White stock. *American Journal of Physical Anthropology* **8**, 23–71.

Tolleth, H. (1987). Concepts for the plastic surgeon from art and sculpture. *Clinical Plastic Surgery* **14**, 4, 585–98.

Tyrell, A. J., Evison, M. P., Chamberlain, A. T. and Green, M. A. (1997). Forensic three-dimensional facial reconstruction: historical review and contemporary developments. *Journal of Forensic Science* **42**, 4, 653–61.

Ubelaker, D. H. (1978). *Human Skeletal Remains: Excavation, Analysis and Interpretation*. Washington DC: Smithsonian Institute Press.

Ubelaker, D. H. and O'Donnell, G. (1992). Computer assisted facial reproduction. *Journal of Forensic Science* **37**, 155–62.

Udry, J. R. (1965). Structural correlates of feminine beauty preferences in Britain and the United States: a comparison. *Sociological and Social Research* **49**, 330.

Valentine, T. and Bruce, V. (1986). The effects of distinctiveness in recognising and classifying faces. *Perception* **15**, 525–35.

Van den Bosch, W. A., Leenders, I. and Mulder, P. (1999). Topographic anatomy of the eyelids and the effects of sex and age. *British Journal of Ophthalmology* **83**, 347–52.

Van der Beek, M. C. J., Hoeksma, J. B. and Prahl-Andersen, B. (1991). Vertical facial growth: a longitudinal study from seven to 14 years of age. *European Journal of Orthodontics* **13**, 202–8.

Vanezis, P., Blowes, R. W., Linney, A. D., Tan, A. C., Richards, R. and Neave, R. (1989). Application of 3-D computer graphics for facial reconstruction and comparison with sculpting techniques. *Forensic Science International* **42**, 69–84.

Vanezis, P., Vanezis, M., McCombe, G. and Niblett, T. (2000). Facial reconstruction using 3-D computer graphics. *Forensic Science International* **108**, 2, 81–95.

Van Rensburg, M. S. J. (1993). Accuracy of recognition of 3-D plastic reconstruction of faces from skulls. Abstract. *Proceedings of the Anatomical Society of South Africa*. 23rd Annual Congress 20.

Viðarsdóttir, U. S. and O'Higgins, P. (2001). Geometric morphometrics and the analysis of variations in facial form: robusticity of biological findings in relation to bilateral versus unilateral and missing landmarks. *Statistica* **2**, 315–33.

Viðarsdóttir, U. S., O'Higgins, P. and Stringer, C. (2002). The development of regionally distinct facial morphologies: a geometric morphometric study of population-specific differences in the growth of the modern human facial skeleton. *Journal of Anatomy* **201**, 211–29.

Virchow, H. (1912). Die anthropologische Untersuchung der Nase. *Zeitschrift für Ethnologie* **44**, 289–337.

Von Eggeling, H. (1909), Anatomische Untersuchungen an den Kopfen von vier Hereros, einem Herero- und einem Hottentottenkind. L. Schultze, Forschungsreise im westlichen und zentralen Südafrika, ausgeführt 1903–1905. *Denkschriften der Medizinisch-naturwissenschaftlichen Gesellschaft zu Jena* **15**, 323–448.

(1913). Die Leistungsfähigkeit physiognomischer Rekonstruktionsversuche auf Grundlage des Schädels. *Archiv für Anthropologie* **12**, 44–7.

Von Hagens, G. (1979). Impregnation of soft biological specimens with thermosetting resins and elastomers. *Anatomical Record* **194**, 247–55.

Wang, D., Qian, G., Zhang, M. and Farkas, L. G. (1997). Differences in horizontal, neoclassical facial canons in Chinese and North

American Caucasian populations. *Aesthetic Plastic Surgery* **21**, 4, 265–9.

Wei B., Feng, J. and Fang, Z. (1983). The relationship between the construction of maxillary first molar and age. *Acta Anthropologica Sinica* **2**, 79.

Weining, W. (1958). Rontgenologische Untersuchungen zur Bestimmung der Weichteildickenmaße des Gesichts. Dissertation Frankfurt.

Welcker, H. (1883). *Schiller's Schädel und Todenmaske nebst Mittheilungen über Schädel und Todenmaske Kants.* Braunschweig: Vieweg und Sohn 1–160.

Wen, I. C. (1934). The development of the upper eyelid of the Chinese with special reference to the mongolic fold. *Chinese Medical Journal* **48**, 1216–27.

White, T. D. and Folkens, P. A. (1991). *Human Osteology.* San Diego: Academic Press Inc.

Whittaker, D. K. (2000). Aging from the dentition. In M. Cox and S. Mays (eds.), *Human Osteology in Archaeology and Forensic Science.* GMM Publishers, pp. 83–100.

Whittaker, D. K. and Richards, D. (1978). Scanning electron microscopy of the neonatal line in human enamel. *Archives of Oral Biology* **23**, 45–50.

Whittaker, D. K., Richards, B. N. and Jones, M. L. (1998). Orthodontic reconstruction in a victim of murder. *British Journal of Orthodontics* **25**, 11–14.

Whittingham, P. D. G. V. (1962). Measurement of tissue thickness by ultrasound. *Aerospace Medicine* **33**, 1121–8.

Wilder, H. H. (1912). The physiognomy of the Indians of Southern New England. *American Anthropologist* **14**, 3, 415–35.

Wilder, H. H. and Wentworth, B. (1918). *Personal Identification.* Boston, MA: Richard Badger, Gormon Press.

Wilkinson C. M. (2002). In vivo facial tissue depth measurements for White British children. *Journal of Forensic Science* **47**, 3, 459–65.

Wilkinson, C. M. and Mautner, S. A. (2003). Measurement of eyeball protrusion and its application in facial reconstruction. *Journal of Forensic Science* **48**, 1, 12–16.

Wilkinson, C. M. and Neave, R. A. H. (2001). Skull reassembly and the implications for forensic facial reconstruction. *Science and Justice* **41**, 3, 5–6.

(2003). The reconstruction of faces showing healed wounds. *Journal of Archaeological Science* **30**, 1343–8.

Wilkinson, C. M. and Whittaker, D. K. (2002). Juvenile forensic facial reconstruction – a detailed accuracy study. *Proceedings of the 10th Meeting of the International Association of Craniofacial Identification*, Bari, Italy 98–110.

Wilkinson, C. M., Motwani, M. and Chiang, E. (2003). The relationship between the soft tissues and the skeletal detail of the mouth. *Journal of Forensic Science*, **48**, 4, 728–32.

Wilkinson, C. M., Neave, R. A. H. and Smith, D. S. (2002). How important to facial reconstruction are the correct ethnic group tissue depths? *Proceedings of the 10th Meeting of the International Association of Craniofacial Identification*, Bari, Italy 111–21.

Williamson, M. A., Nawrocki, S. P. and Rathburn, T. A. (2002). Variation in midfacial tissue thickness of African-American children. *Journal of Forensic Science* **47**, 1, 25–31.

Wolff's Anatomy of the Eye and Orbit, 8th edn. Eds: Bron A. J., Tripathi R. C. and Tripathi B. J. (1997). London: Chapman & Hall Medical.

Yarmey, A. D. (1975). Social-emotional factors in recall and recognition of faces. *Proceedings of Annual Conference of Mid-Western Psychological Association*, Chicago.

Yeatts, R. P. (1992). Measurement of globe position in complex orbital fractures II patient evaluation utilizing a modified exophthalmometer. *Ophthalmic Plastic & Reconstructive Surgery* **8**, 2, 119–25.

Y'Edynak, G. J. and Iscan, M. Y. (1993). Craniofacial evolution and growth. In M. Y. Iscan and R. P. Helmer (eds.), *Forensic Analysis of the Skull*. New York: Wiley Liss Inc., pp. 11–29.

Young, A. W., Hay, D. C. and Ellis, A. W. (1985). The faces that launched a thousand ships: everyday difficulties and errors in recognising people. *British Journal of Psychology* **76**, 495–523.

Zuhrt, R. (1955). Stomatologische Untersuchungen an spätmittelälterlichen Funden von Reckkahn. I. Die Zahnkaries und ihre Folgen. *Deutsche Zahn-, Munds-, und Kieferheilkunds* **25**, 1–15.

Select bibliography

Allison, H. C. *Personal Identification.* Boston, MA: Holbrook Press, 1973.

Angel, J. L. Book Review: Gerasimov's *The Face Finder. Science* **173**, (1971), 712.

Baba, H. Early Jomom female skull from Enshoji, Urawa City, Japan. *Bulletin of the National Science Museum, Tokyo* **18**, 22 December (1992).

Bender, F., Ansen, D. P. and Mihalakis, I. Facial reconstruction of fire victims. Paper in Physical Anthropology section of 34th Annual Meeting of the American Academy of Forensic Science, Orlando, Florida; 8–11 February, 1982.

Borkan, G. A., Hults, D. E., Cardarelli, J. and Burrows, B. A. Comparison of ultrasound and skinfold measurements in assessment of subcutaneous fat and total fatness. *American Journal of Physical Anthropology* **58**, 3, (1982), 307–13.

Briggs, C. A. and Martakis, M. Craniofacial anatomy. In J. G. Clement and D. L. Ranson (eds.), *Craniofacial Identification in Forensic Medicine.* London: Arnold Publications, 1998, 37–48.

Brown, K. A. Developments in craniofacial superimposition for identification. *Journal of Forensic Odonto-Stomatology* **1**, 2, (1983), 57–64.

Bruce, V. Changing faces: visual and non-visual coding processes in face recognition. *British Journal of Psychology* **73**, (1982), 105–16.

Bruce, V., Cowey, A., Ellis, A. W. and Perrett, D. I. *Processing the Facial Image.* Oxford: Oxford Science Publications. 1992.

Charney, M., Snow, C. C. and Rhine, J. S. The 3 faces of Cindy M. Proceedings of the 30th Meeting of the American Academy of Forensic Science, St Louis, MS (1978).

Eisenfeld, J., Mishelevich, D. J., Dann, J. J. and Bell, W. H. Soft–hard tissue correlations and computer drawings for the frontal view. *Angle Orthodontics* **45**, 4, (1975) 267–72.

Gruner, O. Identification of skulls – a historical review and practical applications. In M. Y. Iscan and R. P. Helmer, *Forensic Analysis of the Skull*. New York: Wiley Liss Inc., 1993, pp. 29–45.

Hoffman, B. E., McConathy, O. A., Cavard, M. and Saddler, L. Relationship between the piriform aperture and interalar nasal widths in adult males. *Journal of Forensic Science* **36**, 4, (1991), 1152–61.

Korn, N. and Thompson, F. W. *Human Evolution*. London: Holt, Rinehart and Winston Publishers, 1967.

Lanitis, A., Taylor, C. J. and Cootes, T. F. An automatic face identification system using flexible appearance models. *Proceedings of the British Machine Vision Conference* **1**, (1994), 329–38.

Lomas, D. A canon of deformity; Les demoiselles d'avignon and physical anthropology. *Art History* **10**, 5 (1993).

Mew, J. Use of the 'indicator line' to assess maxillary position. *Cranio-view* October, (1992), 2–24.

Miles, A. E. W. Malformations of the teeth. *Proceedings of the Royal Society of Medicine* **47**, (1954), 817–26.

Phillips, V. M., Rosendorff, S. and Scholtz, H. J. Identification of a suicide victim by facial reconstruction. *Journal of Forensic Odonto-stomatology* **14**, 2, (1996), 34–8.

Prahl-Andersen, B., Ligthelm-Bakker, A., Wattel, E. and Nanda, R. Adolescent growth changes in soft tissue profile. *American Journal of Orthodontists and Dentofacial Orthopedics* **107**, (1995), 476–83.

Richardson, A. and Merrett, J. D. British and West African facial form in ideal occlusion. *Annals of Human Biology* **4**, 4, (1977), 367–74.

Sekharan, P. C. Sexing of skulls via suture pattern types. *Journal of the Forensic Science Society India* **2**, (1924), 19–24.

   Identification of the skull from its suture pattern. *Forensic Science International* **27**, (1985), 205–14.

   Individual characteristics of ectocranial sutures. *Indian Journal of Forensic Science* **1**, (1987), 75–91.

   Permanency of skull suture patterns; evaluation by animal experiments. *Indian Journal of Forensic Science* 1, (1987), 117–24.

Snow, C. C. Victim identification through facial restoration. *Proceedings of the 29th Annual Meeting of the American Academy of Forensic Sciences*, San Diego.

Suzuki, T. Reconstitution of a skull. *International Criminal Police Review* **264**, (1973), 76–85.

Valentine, T. A unified account of the effects of distinctiveness, inversion and race in face recognition. *Quarterly Journal of Experimental Psychology* **43A**, 2, (1991), 161–204.

Yoshino, M., Kubota, S., Matsuda, H., Seta, S. and Miyasaka, S. Face to face video superimposition using three-dimensional physiognomic analysis. *Japanese Journal of Science, Technology and Identification* **1**, 1, (1996), 11–20.

Index